INTRODUCTION TO PSYCHONEUROIMMUNOLOGY

INTRODUCTION TO PSYCHONEUROIMMUNOLOGY

Jorge H. Daruna
Department of Psychiatry and Neurology
Tulane University School of Medicine

ELSEVIER
ACADEMIC
PRESS

AMSTERDAM • BOSTON • HEIDELBERG • LONDON
NEW YORK • OXFORD • PARIS • SAN DIEGO
SAN FRANCISCO • SINGAPORE • SYDNEY • TOKYO
Academic Press is an imprint of Elsevier

Elsevier Academic Press
200 Wheeler Road, 6th Floor, Burlington, MA 01803, USA
525 B Street, Suite 1900, San Diego, California 92101-4495, USA
84 Theobald's Road, London WC1X 8RR, UK

Library of Congress Cataloging-in-Publication Data
Application submitted.

British Library Cataloguing in Publication Data
A catalogue record for this book is available from the British Library.

ISBN: 0-12-203456-2

For all information on all Academic Press publications
visit our Web site at www.books.elsevier.com

Printed in the United States of America
04 05 06 07 08 09   9 8 7 6 5 4 3 2 1

For
Brandon
and
Caroline

PREFACE

This book presents an introduction to psychoneuroimmunology, which is the scientific discipline best poised to elucidate the complex processes that underlie health. It is intended for students planning careers in medicine, nursing, psychology, public health, social work, or for anyone interested in the scientific basis for an integrative approach to healthcare. The book covers a vast territory, touching on various disciplines, including endocrinology, genetics, immunology, medicine, neuroscience, psychology, and sociology. The aim has been to capture the big picture with respect to how bodily systems intercommunicate to sustain health. The book most distinguishes itself from other books on the topic of psychoneuroimmunology by the breadth of coverage, by highlighting the complexity that is evident at all levels of analysis without becoming excessively entangled in the fine detail of specific topics, and by drawing implications for health science research and healthcare.

The book begins with an overview in Chapter 1 of the topics that will be covered in each subsequent chapter. Chapter 2 proceeds to define psychoneuroimmunology within the context of historical ideas about health and disease. The immune system is the topic of Chapter 3. It is described in some detail to highlight that its function is multifaceted and has evolved to protect the organism from diverse microscopic adversaries. However, the immune system does not operate with autonomy. This is made clear in Chapters 4 and 5, which describe how the hormones released by the endocrine system and the activity of the peripheral nervous system modulate immune function. These chapters further underscore that communication between the systems is bidirectional, in other words, the immune system is able to influence the release of hormones as well as the activity of the nervous system. Stress is discussed in Chapter 6 as a broad concept denoting the organism's response to contextual change. Disease is regarded as a form of stress with both physical and psychosocial aspects. Beginning with Chapter 7, the coverage focuses on psychosocial stress and its effects on endocrine, neural, and immune activity. This evidence demonstrates that psychosocial stress can indeed modify the activity of the systems that operate to sustain health. The impact of psychosocial stress on infectious disease, allergy, cancer, and autoimmune disorders is examined in Chapters 8 and 9. It is evident that psychosocial stress interacts with factors such as disease pathophysiology and characteristics of the individual. Its effect is further dependent on temporal factors. Psychiatric disorders are given consideration in Chapter 10, where the role of immune activity in the modification of brain function is discussed. Chapter 11 shifts the focus to factors that appear beneficial with respect to immune function and may serve to safeguard health. The topics covered include the role of suggestion, beliefs and expectations, the influence of social involvement and emotional expression, and the impact of sleep, relaxation, exercise, and diet. Finally, Chapter 12 examines the implications of psychoneuroimmunology and underscores the need for much more integrative approaches to both research and healthcare.

ACKNOWLEDGMENTS

The book has been made possible by the work of numerous scientists and scholars, whose names appear throughout the text. However, my exploration of their work owes directly to the influence of Robert Ader's seminal volume, *Psychoneuroimmunology*. The idea for the book grew out of my experience teaching a course on psychoneuroimmunology at the Tulane University School of Medicine. The course was encouraged by Don Gallant and became possible with the collaboration of Carolyn Daul, Jane Morgan, and Darrenn Hart. Our students reinforced the idea that an introductory book on the topic was needed. I am especially grateful to Martha Ciattei for the inspiration and encouragement, which helped transform the idea into what is now an actual book. Darrenn Hart was involved in the original planning for the book and Mary Watson helped by conducting literature searches. I have enjoyed the support of colleagues, in particular, Richard Dalton, Betty Muller, Dan Winstead, and Charley Zeanah. Paul Rodenhauser offered wise counsel. Patricia Barnes shared her medical knowledge and provided incisive commentary on some of the chapters. I further benefited from the thoughtful comments of anonymous reviewers. I have been most fortunate to receive the superb assistance of Sherry Juul, Julie Aguilar, and Linzi Conners. Julie has been responsible for the majority of the work and has done a magnificent job preparing the manuscript. Arden Nelson provided expert technical assistance. I proudly acknowledge my daughter, Caroline Daruna, for her help with the illustrations. I am greatly indebted to my publisher, Nikki Levy, for her guidance during the unfolding of this project. She and Barbara Makinster have been simply wonderful.

CONTENTS

Preface   vii

Acknowledgements   ix

CHAPTER 1
Introduction

  I. Aim and Organization   1

 II. Source   5

CHAPTER 2
Historical Antecedents

    I. Introduction   7

   II. Health   8

  III. Social Organization, Health, and Healing   9

  IV. Early Ideas about Health and Disease   10

     A. Magic and Religion   10

     B. Natural Philosophy   11

   V. Empirical Approaches to Health   14

  VI. Science and Health   15

     A. Microbes and Innate Defenses   15

     B. Regulation of Life-sustaining Activities   16

     C. Psychoneuroimmunology   18

 VII. Concluding Comments   19

VIII. Sources   20

CHAPTER 3
Immune System Basics

   I. Introduction   23

  II. Molecular Self/Nonself Discrimination   24

 III. Cell Types, Proteins, and Genes   24

   IV. Immune System Cell Types and Complement   25
    V. Immune Cell Functions and Inflammation   28
   VI. Human Leukocyte Antigen System   31
  VII. Antigen Presentation   32
 VIII. Natural Killer Cells   34
   IX. B Lymphocytes, Antibody Structure, and Diversity   35
    X. T Lymphocytes   37
       A. T-Cell Receptor Diversity and CD Markers   37
       B. Helper T Lymphocytes (CD4$^+$)   38
       C. T-Cell Help of Antibody Production   40
       D. T-Cell Help of Cytotoxicity   41
   XI. Cytokines, Chemokines, and Cell Adhesion
       Molecules   42
  XII. Lymphoid Organs and Leukocyte Traffic   46
 XIII. Immune Activation/Deactivation and Memory   49
 XIV. Development of Immune Functions   50
  XV. Measures of Immune Function   51
 XVI. Concluding Comments   53
XVII. Sources   55

CHAPTER 4

Endocrine–Immune Modulation

    I. Introduction   58
   II. Endocrine System   59
       A. Hypothalamus   59
       B. Pituitary Gland   60
       C. Other Glands   62
  III. Cytokines, Hormones, and Their Receptors   63
  IV. Anterior Pituitary Hormones and Immune Function   64
       A. Growth Hormone   64
       B. Prolactin   66
       C. The Pituitary–Adrenal Axis:
          POMC Peptides (ACTH, β-Endorphin)
          and Glucocorticoids   67
       D. The Pituitary–Gonadal Axis: Gonadotropins and
          Gonadal Steroids   70
       E. The Pituitary–Thyroid Axis: Thyrotropin
          and Thyroid Hormones   72

V. Posterior Pituitary Gland Hormones and Immune Function   73
    A. Arginine Vasopressin   73
    B. Oxytocin   73
VI. Other Hormones and Immune Function   74
    A. Insulin   74
    B. Parathyroid Hormone   74
    C. Melatonin   75
VII. Thymus Gland   76
    A. Pituitary Regulation of Thymus   76
    B. Thymus Regulation of Pituitary   76
    C. Hormones and Thymocyte Development   77
VIII. Concluding Comments   77
IX. Sources   79

CHAPTER 5
Neuroimmune Modulation

I. Introduction   82
II. Peripheral Nervous System   82
    A. Somatosensory Pathways   83
    B. Visceral Sensory Pathways   84
    C. Autonomic Nervous System   84
    D. Enteric Nervous System   85
III. Peripheral Nervous System Innervation of
Lymphoid Organs   86
    A. Bone Marrow and Thymus   86
    B. Spleen, Lymph Nodes, and Mucosa–Associated
       Lymphoid Tissue   86
IV. Chemical Signaling in the Periphery   87
    A. Classical Neurotransmitters   87
    B. Neuropeptides   88
    C. Cytokines   88
    D. Other Mediators   89
V. Functional Effects of Peripheral Neuroimmune
Interactions   89
VI. Central Nervous System   92
VII. Bidirectional Central Nervous System–Immune
System Interactions   94
    A. Subcortical Lesions   94

B. Subcortical Responses to Immunization 95
C. Neocortical Lesions 96
VIII. Learning and Immune Responses 98
IX. Personality and Immune Function 99
X. Concluding Comments 101
XI. Sources 102

CHAPTER 6
Stress, Contextual Change, and Disease

I. Introduction 103
II. Selye's Concept of Stress 103
III. Ranking Life Events as Stressful 106
IV. Stress as Contextual Change 108
V. Social Context 109
VI. Other Life Forms 109
VII. Nonliving Environment 110
VIII. Individual as Context 111
IX. Disease as Contextual Change 113
X. Concluding Comments 114
XI. Sources 115

CHAPTER 7
Psychosocial Stress:
Neuroendocrine and Immune Effects

I. Introduction 117
II. Psychosocial Stress 118
A. Life Events 118
B. Individual Attributes/Personality 119
C. Laboratory Paradigms 119
III. Effects on Endocrine Activity 119
A. Pituitary–Adrenal Axis 120
B. Pituitary–Gonadal Axis 121
C. Other Axes and Hormones 121
IV. Effects on Autonomic and Peripheral Neural Activity 122
A. Classical Neurotransmitters 122
B. Neuropeptides 122
V. Effects on the Central Nervous System 123

VI. Effects on the Immune System 123
    A. Nonspecific Immunity 124
    B. Humoral Immunity 125
    C. Cell-Mediated Immunity 126
    D. Cautions and Integration 128
VII. Neuroendocrine–Immune Pathways 129
VIII. Concluding Comments 129
IX. Sources 131

CHAPTER 8

Infection, Allergy, and Psychosocial Stress

I. Introduction 134
II. Infectious Diseases 134
    A. Infectious Agents 135
    B. Pathogenic Mechanisms 137
    C. Infection and Other Diseases 138
    D. Psychosocial Stress, Immunity, and Infection 139
III. Allergic Diseases 143
    A. Allergens 143
    B. Prevalence and Genetics 144
    C. Environmental Cofactors 145
    D. Pathogenic Mechanisms 145
    E. Complexity of Allergic Responses 147
    F. Psychosocial Stress, Immunity, and Allergy 148
IV. Concluding Comments 149
V. Sources 151

CHAPTER 9

Cancer, Autoimmunity, and Psychosocial Stress

I. Introduction 154
II. Cancer 154
    A. Cancer as an Expanding Clone 155
    B. Environmental Carcinogenesis 157
    C. Defenses against Cancer 158
    D. Psychosocial Stress, Immunity, and Cancer 159
    E. Could Psychosocial Factors be Irrelevant? 163

III. Autoimmune Diseases   164
  A. Clonal Selection Theory and Normal Autoimmunity   164
  B. Prevalence of Autoimmune Disorders   165
  C. Pathogenic Mechanisms   165
  D. Infection Triggers Autoimmunity   167
  E. Cancer Triggers Autoimmunity   168
  F. Toxic Chemicals Trigger Autoimmunity   168
  G. Susceptibility to Autoimmunity   169
  H. Psychosocial Stress and Autoimmune Disorder   170
  I. Gender and Autoimmunity   174
IV. Concluding Comments   175
 V. Sources   177

CHAPTER 10
Immune Activity and Psychopathology

  I. Introduction   180
 II. Access to Brain by Pathogens   181
III. Immune Activity within the Brain   181
  A. Lymphocyte Entry   182
  B. Cytokine Effects in Brain   182
  C. Optimal Immune Response in Brain   183
  D. Consequences of Immune Activity in the Brain   183
IV. Nervous System Infections and Behavior   184
  A. Neurological Disorders   185
  B. Neurodegenerative Disorders   186
  C. Lyme Disease   187
  D. Herpes Viruses   187
  E. Rabies Virus   189
  F. Human Immunodeficiency Virus   189
  V. Autoimmunity, Malignancy, and Behavior   190
  A. Systemic Lupus Erythematosus   190
  B. Paraneoplastic Disorders   191
 VI. Sensing Peripheral Immune Activity   191
  A. Cytokines Alter Mental Processes and Behavior   191
  B. Channels of Communication   192
  C. Functional Significance   193
VII. Sickness Behavior   193
VIII. Behavioral Disorders that Resemble Sickness Behavior   194
  A. Chronic Fatigue and Pain   194
  B. Depressive Disorders   195

IX. Psychiatric Disorders with a Link to Infection   196
     A. Autism and Pervasive Developmental Disorders   196
     B. Attention-Deficit/Hyperactivity Disorder   197
     C. Childhood Obsessive-Compulsive Disorder
        and Tourette's Syndrome   198
     D. Schizophrenia and Other Psychoses   199
  X. Concluding Comments   203
XI. Sources   204

CHAPTER 11
Immune Function Enhancement

   I. Introduction   207
   II. Beliefs, Suggestion, and Expectations   208
     A. Hypnosis   209
     B. Placebo Effect   210
     C. Nocebo Effect   211
     D. Expectations are Patterns of Neural Activity   211
     E. Expectation and Immune Activity   212
  III. Social Engagement   213
  IV. Expression of Emotion   213
   V. Sleep and Relaxation   214
  VI. Exercise and Physical Activity   215
     A. Leukocytes   215
     B. Acute-Phase Proteins, Antibodies, and Cytokines   216
  VII. Nutrition   217
     A. Malnutrition   218
     B. Individual Factors and Nonlinear Effects   218
     C. Antioxidants: Vitamins and Minerals   219
     D. Lipids   220
     E. Modulation by Hormones   220
VIII. Concluding Comments   221
  IX. Sources   222

CHAPTER 12
Integration and Implications

   I. Introduction   225
   II. Synopsis   225
  III. Microenvironments and Complexity   228

      IV. Unnecessary and Insufficient   228
       V. Implications for Research   229
            A. Limits of Reductionism   229
            B. Major Dimensions of Health State Space   229
      VI. Implications for Health Care   232
            A. Holistic-Expanded Perspective   232
            B. Individualized Health Education   232
     VII. Economic Considerations   233
    VIII. Concluding Comments   233
      IX. Sources   234

      Glossary   235
      Index   261

CHAPTER 1

Introduction

I. Aim and Organization   1
II. Source   5

I. AIM AND ORGANIZATION

The term *psychoneuroimmunology* was first introduced by Robert Ader during his presidential lecture to the American Psychosomatic Society in 1980. In that lecture, he summarized research that demonstrates the fundamental unity of the bodily systems that function to maintain health, and he underscored the fact that the immune system is no exception to this general rule.

This book provides an overview of the field of psychoneuroimmunology as it has taken form over the last 25 years. While reading, keep in mind that what follows is a series of sketches taken from one observer's perspective. The intent is to make psychoneuroimmunology more accessible to a wide readership without oversimplifying or becoming so entangled in much of the available detail that the big picture is lost.

The relevant literatures are vast and cut across many disciplines, so an effort such as this runs the risk of overlooking some relevant findings or overemphasizing observations that have been widely reported. Nonetheless, it seems unlikely that such possibilities would invalidate the overall picture that this effort aims to paint. Thus, the presentation is not exhaustive and often relies on secondary sources.

Given that an important theme of this book is the overriding impact of context at all levels, it seems fitting to begin by placing psychoneuroimmunology in historical context, specifically with respect to the evolution of ideas about health and disease. Consideration of historical antecedents in Chapter 2 underscores that ancient medical wisdom already recognized much of what modern science has come to discover as beneficial. Indeed, it may be fair to say that even though knowledge of the intricacies of nature has grown tremendously, the fundamental dynamics at play in nature have long been recognized. Therefore, psychoneuroimmunology constitutes a scientific demonstration of the long-recognized unity of the organism.

The fundamental unity of the organism is the key notion emphasized throughout this book. However, the scientific understanding of unity has benefited from the deconstruction of organisms into interacting systems. Therefore, in

1

Chapter 3, the presentation moves to a relatively detailed description of the immune system, though at a basic level and without delving any more than necessary into the molecular realm. Essentially, the immune system is presented as a collection of cell types and large molecules whose activity serves to maintain organismic integrity by neutralizing pathogens. The level of detail in this chapter underscores the difficulty one encounters in trying to globally characterize immune function. Moreover, it is important to recognize that immune competence is variable across individuals, phases of development, and contexts. Finally, methods of quantifying immune activity are described, making it apparent that they constitute rather limited windows into the operation of the immune system.

The immune system is poised at the interface between the organism and microscopic adversaries. Its activity is crucial to the integrity of the organism. However, it does not operate with autonomy. Chapter 4 focuses on hormones and peptides released by endocrine glands that have documented effects on the immune system. The intent is to gain insight into the role of the endocrine system in the modulation of immune function. Again, details are necessary to do justice to the complexity of the interactions that must be taken into consideration. However, the key idea to keep in mind is that endocrine hormones are central to the mobilization and storage of energy resources and to the growth and development of the organism leading ultimately to the generation of new life. This chapter begins the process of viewing the immune system as operating within the context of other bodily systems, which intercommunicate via molecular signals.

The topic of intercommunication is further explored in Chapter 5, which deals with the modulation of the immune system by the central nervous system (CNS). Nervous system activity is in part transmitted to the immune system via the endocrine system, as described in Chapter 4. Neural signals also act directly on immune cells in various tissue compartments through the peripheral nervous system innervation of specific tissues and organs. In turn, immune activity in the periphery is capable of initiating neural impulses that are conducted on all levels within the CNS. The CNS is pivotal with respect to adaptive function and so this interconnectivity further underscores the fundamental unity of the organism. The task of survival in a changing environment depends on the brain, the immune system, and their intercommunication.

Stress is the subject of Chapter 6. Originally, the concept of stress stems from the observation that organisms exposed to a variety of noxious conditions, including infections, toxins, drugs, extreme temperature, electric shock, or physical restraint, responded with a characteristic pattern of effects on immune and endocrine tissues. It was further observed that the effect of any such challenge was dependent both on whether the organism had been previously challenged and on the time interval between challenges. It should be noted that the range of conditions capable of producing stress is rather broad and that the organism's response to stress depends on its preexisting state, which implies that all had transpired in the past including what is coded in the genes. The perspective espoused in this chapter is that stress is fundamentally a response to contextual change and that clearly the direction of its effect is dependent on the preexisting state of the organism. Stress can have adverse and beneficial effects.

The topic of Chapter 7 concerns how psychosocial stress has an impact on endocrine and immune activity. Psychosocial stress is a form of contextual change arising either from disruptions in relationships or social networks or as a result of demands for personal performance with implications for well-being. In essence, the aim of this chapter is to examine the human evidence lending support to the interconnectivity between the systems described in the preceding chapters. Changes in the psychosocial context of individuals are assumed to alter brain activity, and to the extent that alterations in endocrine and immune activity are also evident, support is gained for the role of psychosocial stress in the modulation of immune functions. The question of whether such changes affect the risk of disease is examined in Chapters 8 and 9.

Chapters 8 and 9 present overviews of major disease categories including infectious disease, allergy, cancer, and autoimmunity. These disorders have in common the involvement of immune activity. They are all considered to be forms of stress. Thus, the impact of psychosocial stress should be dependent on its timing and on the way in which the immune system participates in the disease process.

The discussion of infectious disease examines how infections spread and emphasizes the fact that infection is not tantamount to disease. Moreover, there are individual differences in clinical symptoms, their severity, and how the illness progresses in response to a given pathogen. Psychosocial stress appears as an additional modifier, whose impact depends on characteristics of the individual, the infectious agent, and temporal factors.

Allergic disorders are presented as instances of immune reactivity to innocuous external material. However, there is now persuasive evidence that frequent treatment of early infections with antibiotics and generally more hygienic environments in early life may actually increase the risk of allergy. Psychosocial stress appears capable of both augmenting and attenuating allergic responses. Particularly noteworthy is that allergic responses can be conditioned and are highly susceptible to suggestion.

Cancer arises as a result of accumulated failures in the genetic regulation of cell proliferation. The immune system appears to play a role, at least for some forms of cancer. Psychosocial stress can have an impact on the genesis and progression of cancer. This can occur via multiple pathways including the immune system, but the effect may be modest in magnitude and dependent on the specific type of malignancy.

Autoimmune diseases are portrayed as arising out of dysregulation of immune mechanisms that inhibit responses to self-antigens that are part of healthy tissue. Psychosocial stress appears to have an impact on the occurrence and progression of these disorders. Again, the magnitude and perhaps even the direction of the effect may be variable across specific disorders and dependent on individual characteristics.

Overall, psychosocial stress emerges as a factor with respect to disease onset and progression. However, it is not simply a factor that brings about or accelerates disease. There are instances in which psychosocial stress may protect against or decelerate disease progression. Individual characteristics, the nature of the underlying pathophysiology of the diseases, and temporal relationships are all factors that modify the outcome.

The role of infection and immune activity on cognition, emotion, and behavior is the topic of Chapter 10. How pathogens influence brain activity is examined. There are many documented instances in which behavioral disturbances are at least in part attributable to infection of the brain or to immune activity triggered in the periphery. There are also disorders of activity and emotion that resemble the changes in demeanor, known as *sickness behavior*, which accompany the acute-phase response to infection. Involvement of pathogens or of immune activity in some instances of behavioral disorders is not evidence that they should be expected in every case. The importance of this evidence is that it emphasizes the need not to overlook the possibility of infection in the etiology of behavioral disorders, ranging from schizophrenia to attention-deficit disorder or from obsessive-compulsive disorder to depression and even to conditions such as chronic fatigue syndrome. The role of infection in psychiatric disturbance has probably been underestimated.

Chapter 11 considers factors that appear to enhance immune function. Hypnotic suggestion and other procedures that induce expectations are discussed. The effects of hypnosis, placebo, and nocebo are clear instances of how expectations generated in the context of a relationship influence bodily process. It is emphasized that this is a natural consequence of the fact that expectations are central patterns of neural activity that can have peripheral effects. The evidence further documents that social engagement and the communication of troubling experiences associated with negative emotions are beneficial to immune function and health. The contributions of sleep, relaxation, diet, and exercise are also considered.

Having examined the major facets of psychoneuroimmunology in the preceding chapters, Chapter 12 attempts to create an integrated picture of the field and draw implications for research and health care. The effort at integration forces a focus on microenvironments within the organism as the likely site in which health is maintained or the process of disease is initiated. A multitude of variables interact within microenvironments and what occurs from moment to moment does not appear to depend on any single variable. Disease emerges as a result of complex interactions among many variables. Specific variables are neither necessary nor sufficient as a general rule. Moreover, variables can be categorized as serving to increase or decrease risk of disease. The same variable may increase or decrease risk for a given disease depending on when it occurs with respect to other variables. In addition, a variable that increases the risk of a particular disease may also decrease the risk of another disease. Therefore, what matters is not the presence of specific variables, but the occurrence of patterns or configurations of variables including those that define the disease itself. The implications for research are clear; the focus should be on uncovering the patterns, which requires simultaneous measurement of a wide range of variables. In turn, clinical care should be guided by the fact that disease is a multidimensional condition that is not adequately confronted using one-dimensional treatments. In closing, consideration is given to the economic climate in health care and how it may have an impact on practices guided by the integrative perspective embodied in psychoneuroimmunology.

II. SOURCE

Ader, R. (1980), "Psychosomatic and Psychoimmunologic Research," *Psychosomatic Medicine*, 42, pp. 307–321.

CHAPTER 2

Historical Antecedents

I. Introduction   7
II. Health   8
III. Social Organization, Health, and Healing   9
IV. Early Ideas about Health and Disease   10
 A. Magic and Religion   10
 B. Natural Philosophy   11
  1. Chinese Ideology   12
  2. Indian Ideology   12
  3. Greek Ideology   13
V. Empirical Approaches to Health   14
VI. Science and Health   15
 A. Microbes and Innate Defenses   15
 B. Regulation of Life-sustaining Activities   16
 C. Psychoneuroimmunology   18
VII. Concluding Comments   19
VIII. Sources   20

I. INTRODUCTION

Psychoneuroimmunology is essentially an integrative discipline. It seeks to shed light on how mental events and processes modulate the function of the immune system and how, in turn, immunological activity is capable of altering the function of the mind. Psychoneuroimmunology encompasses a sufficiently broad area of scientific research, from the molecular to the interpersonal, so it has the potential to lead to the development of a more comprehensive model of health. This chapter begins by considering the notion of health and how health has been understood historically. This historical perspective will serve to put psychoneuroimmunology in the context of the evolution of thought concerning health.

II. HEALTH

Health is a characteristic of life, which may be evident in the appearance and behavior of organisms. In the case of social organisms, changes in the health of one individual can affect the behavior of others. For instance, other individuals may change their movements to remain near the affected one. They may bring food to the sick member and give other forms of active assistance. These observations have been made frequently in the case of chimpanzees living in the wild and are probably detectable in other species living in groups.

Evidently, the misfortune of one organism affects the group by eliciting, at least temporarily, helpful behaviors. Postural changes, activity levels, and emotional displays all serve to transmit the health status of the individual. Typically, in natural settings biologically related group members tend to be more responsive to one another's health status. However, biological relatedness is a matter of degree and may extend to familiarity, because in some circumstances such as captivity, animals will give assistance and support to unrelated group members or members of other species.

Fábrega (1997) has theorized that the behavioral responses to changes in health by individuals and their immediate social group are the precursors to what gradually has become the healing enterprise of humanity. He assumes that long before the dawn of human civilization, early humans, most likely family-level foragers, would at least have shown the level of health responsiveness evident in nonhuman primates. For these early humans, many of their activities, during their modest life span, were geared toward health maintenance in the form of basic survival. One component of this survival orientation was to give care to temporarily disabled group members. However, *temporary* had to be a key element in the sense that if signs of recovery were not evident, group behavior shifted as though in anticipation of death. Even temporary assistance would not always be adaptive to the individual giving it and would ultimately hinge on the balance between costs and possible benefits to the group resulting from the survival of the afflicted individual. Thus, it appears likely that early humans operated according to the dictum "risk the costs of short-term care to promote recovery, but not simply to delay death," particularly in the case of those individuals who could contribute to group well-being in the future.

This basic adaptation of giving calculated aid, as well as seeking it should the need arise, underscores the social nature of health maintenance. Fabrega (1997) has argued that there is a biological disposition to seek and give care when health becomes compromised. This disposition is progressively transformed with changes in social organization and cultural evolution into what can only be described as the health care industry. Sociocultural changes not only have an impact on health itself, but also affect how diseases are understood and labeled and how and by whom they are treated for what sort of return.

III. SOCIAL ORGANIZATION, HEALTH, AND HEALING

Ideas about health and approaches to healing have gradually evolved in relationship to the organization of individuals into social structures of various levels of complexity. Fábrega (1997) has written on the likely progression of this process. According to his scheme, family-level foragers did not possess elaborate causal explanations of sickness. In general, for them causes were few and concrete. In the case of persistent sickness, spirits or enemies may have been invoked in attempts at explanation.

Village-level communities appeared as the ability to cultivate the land and exploit localized resources developed. The conditions of life in such communities resulted in more sickness and saw the advent of specialized healers (previously family members performed the healing functions). Loss of health was seen as before, although moral and ethical considerations began to be added to the list of causes for sickness.

Gradually, chiefdoms or state societies arose as the more populated regions became politically organized and socially differentiated. There was more pervasive sickness, and healing emerged as an occupation. Life in these communities was characterized by more social conflict, rivalries, and jealousies, which were thought to affect health by acting through the practices of magical rituals conducted for the purpose of causing harm. Causation of sickness still was very much focused on religion, morality, and cosmology. The notion that sickness "just happened" from obvious natural causes also remained as an explanation. At the same time, there was growing awareness that the healthy state implied a sense of connection of the individual to family, group, and setting. Disruption of this harmony brought on sickness.

Eventually, full-fledged civilizations and empires came into existence. These were complex societies with highly elaborated cultures. The burden of disease was great, and chronic diseases, also referred to as *diseases of civilization* (e.g., diabetes, coronary heart disease, hypertension, and cancer), became more prevalent. Psychiatric disorders were more evident. Socioeconomic factors were noted to affect risk of sickness. Exploitation by healers began to be a concern and gave rise to ethical codes of conduct for healers. The classic medical traditions of antiquity were born in China, India, and the Mediterranean. These were holistic orientations to sickness and healing. They emphasized that health required the proper balance of multiple factors; and despite differences in the specific details, the major traditions coincided at a more abstract level. Balance of elements and processes was central to all health. Multicausality and holism guided attempts at healing.

Ideas about health underwent little change until the influence of science began to be felt in the understanding of disease. The book by Golub (1994) on the limits of medicine is particularly enlightening in this regard. He notes that even before the influence of microbes had been discovered, observations about the patterns of infectious diseases indicated that sickness could be affected by living conditions and resulted in societal changes that promoted sanitation. It was not until the mid-nineteenth century, in modern European society, that the role of germs was discovered. The germ theory of disease launched the biomedical approach to sickness, which was in contrast to what had been advocated by the classic traditions. Medical expertise became fully secular and cut off from spiritual, religious, moral, or cosmological overtones. The physician became the health expert in a position of dominance with respect to the patient, although there were dissenting voices, which gave rise to natural healing movements (e.g., homeopathy, osteopathy, and naturopathy).

As scientific advances have brought into focus the functional complexity of the systems within the body, which serve to maintain health, the limitations of biomedical reductionism have become apparent. Interest in alternative, unorthodox, Eastern, and holistic healing practices has regained status. Moreover, the role of social factors and psychological outlook has been given scientific validation with respect to influencing the probability that illness will occur. These developments have coincided with economic concerns over health care expenditures that are redefining how individuals gain access to expertise and treatment. The bureaucratic and impersonal approach to sickness that has become prominent in the current cost-cutting environment has diminished trust in healers. Consequently, there is increasing individual preoccupation with sustaining health. Exploitation of the fear of disease has become a multibillion dollar industry. Over-the-counter pharmaceuticals and health-promoting products are more diverse and readily available than ever before.

IV. EARLY IDEAS ABOUT HEALTH AND DISEASE

A. MAGIC AND RELIGION

Some of the earliest records of medical thought date back to between 3000 and 2500 B.C. The content of such documents is certainly older and can be most readily traced to Egyptian, Babylonian, Hindu, and Chinese civilizations that had come into existence more than 6000 years ago. In the ancient world, magic, religion, and medicine were undifferentiated. Disease could be the result of some everyday event with no further explanation needed. However, it could also occur as a result of the will of divine beings as punishment for sin. Magic invoked by other humans could cause disease.

Illness could be brought about by directing evil intent at some representation of the person or some item belonging to the person. Such representation of the person or item was thought to give access to the individual's soul. Ill will could be transmitted by making eye contact, as in the practice known as "evil eye." In essence, deities made mortals sick as punishment either directly or through the action of intermediary demons. Other humans could also cause disease through the invocation of curses. Given these presumed causes of disease, which included the sins of parents in the case of childhood afflictions, it followed that the treatments consisted of placating the irate deities, driving out the demons, or fighting the curses. Rituals, prayers, incantations, and potions were typical remedies to combat the various afflictions.

Such ideas were predominant over the period up to the latter part of the first millennium B.C. They recurred in the Christian world beginning from the fourth century A.D. until the Renaissance, when more rational views gained prominence again. It should be noted that even in the heyday of magical/religious medicine, less supernatural ideas were also in vogue. For instance, in ancient Egypt, the notion was advanced that disease resulted from problems with the function of a system of vessels originating in the heart and connecting it to all parts of the body. These vessels carried the substances of the body, all thought to originate from the heart, which was considered to be the central organ, the seat of feeling, thinking, and the activities of all the other organs. Cosmic influences were thought to act directly on the humors. Just as the moon affects the tides, the celestial bodies were thought to influence the flow of the humors. The importance of avoiding contagion was also recognized in the form of regulations regarding burial of the dead, disposal of waste, and sexual contact.

It is evident from this brief overview of early ideas about disease that religious/magical and naturalistic views coexisted. Causation was often attributed to deities, thought to be the major forces operating to bring about life in the world. The importance of social relationships and personal conduct was recognized in the role of ill will, magic, and sin. The importance of infection of the body as a cause of disease was evident in the ideas of demonic intrusion. Science has dramatically changed how disease is understood, but it should be noted that religious/magical notions are still invoked in some quarters of our postmodern world.

B. NATURAL PHILOSOPHY

These approaches to understanding health were part of natural philosophical outlooks regarding the operation of the universe and, therefore, of its constituents including human beings. The written sources of this information that have been uncovered date from as early as the latter part of the first millennium B.C. (i.e., 500–200 B.C.). Major traditions come from the Chinese, the Indian, and the Mediterranean (Greek) civilizations. These approaches to conceptualizing health and its disruption are elaborate and complex, but in essence they all rely on the notion that substances flowing within the body must do so in harmony and unimpeded for health to exist. Disease occurs when the flows are somehow disrupted and imbalanced. These similarities across cultures may be evidence of intercommunication or may suggest that confronted with the same phenomena (e.g., body structure, sickness, and death), similar explanatory theories would necessarily emerge.

1. Chinese Ideology

The Chinese tradition dates back more than 4500 years. It is rooted in Taoist ideology, which viewed the body as a replica of the universe. It has an internal landscape that is governed by *xin*, translated "heart-mind." Change and movement are the essence of the universe, as it is for the body. This is represented in the body by the flow of *qi*, not easily translated into English but intended to signify that which must be present in the right proportion at the various locations in every moment to sustain health. Imbalances can be described in terms of two abstract qualities, *yin* and *yang*, which can be used to represent polar opposites, such as *cold* and *hot*.

Qi performs key bodily functions. For instance, deficient protective *qi* (*zhengqi*) makes one susceptible to illness because one is unable to cope with changing circumstances. Such changes can arise from external sources (e.g., climate) or internal sources such as excessive or insufficient emotion (*qiqing*). Again disease results from disharmony; the *qi* has been blocked somewhere in its circulation and lacks the proper spatiotemporal *yin-yang* configuration. Treatment aims to redistribute *qi* properly over the entire body. This ultimately leads to restoring *jing* (tranquility) in a person's *xin* (heart-mind). In essence, to keep one's health, one must guard one's *jing*. This is accomplished in a preventive way by conducting a proper life (nutrition, exercise, and the correct attitude toward one's surroundings). When disease occurs, restoration of the original balance is the goal, and to this end, herbal extracts, massage, acupuncture, and other remedies are employed. These are thought to work by strengthening the body's capacities for curing itself.

2. Indian Ideology

The Indian tradition, *Ayurveda* (knowledge of longevity) dates back more than 4000 years. It is rooted in Sankhya philosophy, which postulates that desire for existence (life) caused the beginning of the universe. Mind exists first, it forms body, and then mind is dependent on body. Mind and body are inseparable. The mind/soul is fixed in the body and depends on the body's health for the quality of its expression. At the same time, perversity of mind is the ultimate cause of every disease.

The state of mind is depicted as reflective of the balance between three principles, referred to as *dosas*. They are *sattva* (the mind's natural state), *tamas* (the focus on external objects), and *rajas* (the focus on action). Predominance of the latter two suppresses the former, causing the mind to function inefficiently. However, it is through action on the body's *dosas* that illness ultimately occurs. The body's *dosas* are *vāta*, *pitta*, and *kapha*. They are active and difficult-to-control condensations of the five great elements of nature (i.e., air, space, fire, water, and earth). All three *dosas* must work together to sustain health. *Vāta* directs all motion, *pitta* is in charge of transformations, and *kapha* is the stabilizing influence. When appropriately balanced, the *dosas* cause the body's elements to cohere and function together healthily. *Dosas* have specific qualities (e.g., *vāta* is dry, cold, light . . . , *pitta* is moist, hot, light . . . , and *kapha* is wet, cold, heavy . . .). All substances and activities in the universe increase or decrease the *dosas*. For instance, according to the law of like and unlike, hot substances increase *pitta* but decrease *vāta* and *kapha*. Experiences must be "digested" (assimilated) by the digestive fire (*ãgni*), which is kept strong by well-balance *dosas*. The balance or imbalance of the *dosas* creates health or disease.

Imbalance of the *dosas* can be avoided by adhering to a daily and seasonal routine of activities. These practices should be especially observed during pregnancy to ensure that the infant is given the proper start. Individuals differ in the predominance of specific *dosas*, which gives them a different temperament or *parakritic*. This must be taken into consideration when providing treatment to reestablish balance. Treatments rely on stimulation of the five senses (massage, aromas, music, etc.), purification (e.g., enema and emesis), and medicines including not only substances but also "appropriate" thoughts—all aimed at balancing the *dosas*.

3. Greek Ideology

The Mediterranean tradition, better known as *Greek medicine*, is associated with the teachings of Hippocrates (early part of fourth century B.C.) and his followers, the most renown being Galen. According to some scholars, the flowering of medical thought in Greece was relatively rapid and had to have benefited from the ancient traditions probably transmitted via Persia. The ancient traditions of medicine, coupled with the Greek emphasis on free thought, catapulted Greek medical thought to a position of prominence. Within the Greek tradition, the balance of four humors is central to health. The humors were blood, phlegm, yellow bile, and black bile. They were basic in the same way that nature was thought to be composed of four basic elements (air, water, fire, and earth). The humors possessed different qualities of temperature (hot or cold) or moisture (dry and wet). Each humor had its origin in a different part of the body (e.g., phlegm in head, blood in heart, yellow bile in gall bladder, and black bile in spleen). Humors were constantly renewed from food and water and could be transformed into one another. Disease resulted when there was imbalance, either excess or insufficiency of one or more humors. For instance, excessive black bile was thought to cause melancholia and malignancies. Humors were generated, kept in motion, blended, and transformed by the action of *innate heat*, the body's sacred fire, which was thought to originate in the left ventricle of the heart.

The Greeks had noted that illnesses could be seasonal and that some types of individuals appeared more susceptible than others. They incorporated these observations within a framework that related humors to seasons and postulated that the humors gave rise to temperaments. Again, disease was due to an upset balance of the humors, and the humors were everywhere in the body. This explained why in every case the whole individual was sick and not just a part. Therefore, treatment was to be not only local, but also general, and the importance of the psychological was not overlooked. The goal was to reestablish the balance of humors by mobilizing the body's natural healing processes. The physician was there to assist the natural healing process. He was nature's servant. If nature resisted, all measures were in vain. As in the other traditions, preservation of health was the priority. Health depended on a proper way of life. This included suitable food, drink, activity, living conditions, exercise, sleep, sexual activity, and psychic activity. Those who became passionate or excited, even for the most unimportant reason, frequently became ill, and it was difficult to heal them.

The Chinese, Indian, and Greek traditions, despite superficial differences, are consistent in (1) recognizing health as the result of a balance of substances, (2) conceptualizing the individual as integrated in a most fundamental way, (3) emphasizing prevention as the goal of medical knowledge, and (4) recognizing the organism's ability to self-heal, which is what treatments attempt to enhance when illness is present. With regards to treatment, it appears evident that efforts were global for the most part, and although possessing a superficial logic (manipulations such as bloodletting to affect the flows and levels of the humors or other essences), there was very little scrutiny of treatment effectiveness. Perhaps having treatments may have been more important than whether they had efficacy. Indeed, treatments such as bloodletting and emesis could have in some cases accelerated a patient's demise.

V. EMPIRICAL APPROACHES TO HEALTH

The classic medical traditions were well developed by the early part of the first millennium. Nonetheless, people died young. Golub (1994) has pointed out that until relatively recent times, about 25% of children died during the first year of life. Another 25% died before the age of 20 years, 75% had died before age 45, and fewer than 10% reached age 60. Human experience with plagues had made it apparent that crowded living conditions and poor sanitation were frequently associated with such epidemics. The idea that contact promoted the spread of disease was recognized by the Egyptians and had continued to be emphasized but did not always promote practices that could be effective. For instance, the dominance of religious explanations suggested that mass appeals to God for mercy could help. This required large groups to congregate in proximity, thus, actually facilitating spread of disease.

The circumstances that promoted rampant spread of disease did not begin to change appreciably until well into the nineteenth century. Disease was still primarily thought of in terms of imbalance of the humors, but by 1840 the sanitation movement was driven by the view that filth caused imbalance of the humors and had as its mission to eliminate disease by getting rid of dirt. Improvements in the living conditions of the poor were justified, not only for their benefit but also to safeguard the property and security of the affluent. In England, this movement ultimately led to the passing of the Public Health Act of 1848. The changes led to dramatic decreases in mortality, even though the guiding understanding of disease (imbalance of the humors) had not yet changed appreciably.

VI. SCIENCE AND HEALTH

Developments in ideology, which would ultimately lead to the flowering of science, are said to have been launched during the period of European history known as the Renaissance. Science as we know it today began in the seventeenth century. However, during the early history of the scientific enterprise, there was very little impact on how health and healing were viewed, even though important steps were being taken. In 1543, Adreas Vesalius published a major human anatomy book. William Harvey published research on the circulation of blood in 1628. A major technological development was the invention of the microscope by Anton van Leeuwenhoek in 1670. With this device, he went on to discover the existence of microscopic life, which he called "animacules." Nonetheless, the approach to illness in the eighteenth century remained very much like what had been practiced for more than 15 centuries.

A. MICROBES AND INNATE DEFENSES

This situation began to change toward the middle of the nineteenth century. Credit for beginning the changes in how disease was understood goes to Pasteur. He is recognized as the first person to demonstrate that microscopic organisms (i.e., bacteria) were capable of causing chemical reactions that could be thought of as "disease." His discoveries led to the germ theory of disease, which was in turn bolstered by Lister's demonstration that sterilization reduced sepsis associated with surgery. Others had begun to show the role of bacteria in diseases of both humans and animals. The discoveries were beginning to make it clear that infectious diseases were specific and transmittable in unique ways. By the 1880s Pasteur's work had set in motion the idea that there could be specific prevention of disease. This was not a new idea. It had been evident since Edward Jenner realized that cowpox infection (causing a mild disease in humans) protected individuals from contracting smallpox (a potentially lethal disease). Already by 1798, Jenner had established vaccination (*vacca* is Latin for *cow*) as protection against smallpox. However, even though the efficacy of vaccination had been shown, during the early 1800s there was enough negative reaction to the idea of infecting people to protect them that antivaccination societies came into existence. Some apprehension about the negative effects of vaccination remains to this day. Discoveries continued to be made that underscored the observation that small amounts or weakened strains of microbes or their products (i.e., toxins produced by bacteria such as those causing diphtheria or tetanus) made individuals resistant to particular diseases, but not to others. Emil Behering and Shibaraburo Kitusata were instrumental in bringing about the idea that "something in the blood" served to protect the vaccinated individual.

Elie Metchnikoff is given credit for being the first investigator to recognize that the cells seen at sites of inflammation were the body's way of combating the infecting bacteria. He put forth this view in 1883 and has come to be regarded as the founder of immunology, specifically for his recognition that the body has an active mechanism of defense. The cells observed at the site of infection were ingesting the bacteria and were called *phagocytes* (i.e., cells that eat). Others, including Almroth Wright, became convinced that vaccination worked by stimulating the phagocytes so that they were better able to dispose of specific pathogens. How this might work in the case of toxins released by bacteria or other infectious agents was first proposed by Paul Ehrlich in 1900. He had done much work on the specificity of dyes for staining different types of tissues. He reasoned that such specificity implies that something on the tissue, which he termed a *side chain*, specifically binds to some dyes but not to others. He further reasoned that if this was true for dyes, why not for toxins. In other words, toxins would preferentially attach to some side chain with an appropriate configuration. This somehow would result in more side chains being produced and shed into the circulation so they would constitute the protection in the blood, which had been postulated by Behring and Kitusata. Thus, by the beginning of the twentieth century, the immune system had begun to be characterized.

B. REGULATION OF LIFE-SUSTAINING ACTIVITIES

During the latter half of the nineteenth century, the work of Claude Bernard had demonstrated that the internal environment (i.e., *milieu interieur*) was maintained in a balanced state. According to Bernard, "The balance of chemicals between the tissues of the body is what determines health." Clearly, the idea of health as an expression balance remained in the picture even as science was becoming the dominant approach to understanding health and disease. More recently, Golub (1994) has concluded that "the biology of complexity will cause us to return to understanding that health really is a form of internal balance," a view that is further emphasized by Sternberg (2000) in her book *The Balance Within: The Science Connecting Health and Emotions.*

The early twentieth century also saw developments in understanding the endocrine glands and the nervous system. The study of the nervous system had been progressing in parallel to that of infectious disease. The first empirical demonstration of brain control over a vital function for life was the discovery of the respiratory center in the medulla oblongata (brainstem) by Jean-César Legallois (1806). However, observations dating back to the mid-sixteenth century and again repeated by Willis (1664) indicated that there was a group of nerves that affected the movement of the heart. These connections were thought, as had been previously theorized by Galen, to permit physiological "sympathy," that is, functional unity or harmonious communication between internal organs involved in the circulation and respiration. In 1732, Jacobus Winslow introduced the notion of the "great sympathetic nerves," which he believed controlled the viscera. By 1845, Weber and Weber had shown that stimulation of one of these nerves (the vagus nerve) could stop the heart, whereas stimulation of the other nerves accelerated the heart rate. These observations led Walter H. Gaskill (1886) to the conclusion that the "involuntary system" was composed of two antagonistic components. John Newport Langley (1898) named this system the *autonomic nervous system* (ANS) and referred to its antagonistic components as the *sympathetic* and *parasympathetic branches.*

The word *hormone*, with Greek roots meaning "to arouse," was first used in 1905 by Starling in the lecture "The Chemical Correlation of the Function of the Body." An understanding of the powerful influence of endocrine gland secretions had been steadily growing since the 1830s. By the mid-1940s many hormones had been chemically characterized and were thought to exert their effects by acting through receptor molecules. The role of higher order brain centers had become the focus of investigation as well. Walter Cannon concluded that the hypothalamus regulated ANS responses directed toward mobilizing bodily resources and preserving life under challenging conditions. A direct connection between the hypothalamus and the ANS was first demonstrated by Beattie, Brow, and Long (1930). Stimulation of this pathway accelerated heart rate, mobilized the body's energy reserve (i.e., increased glucose in the blood), and raised core body temperature (i.e., produced fever). Regulation of pituitary hormone release by the hypothalamus was eventually demonstrated in the late 1960s.

From the end of the nineteenth century to the beginning of the twentieth century, Ivan Pavlov was conducting experiments that demonstrated that visceral responses (e.g., salivation and gastric motility) could be elicited by neutral stimuli (e.g., sound of bell) through a process of temporal pairing of the neutral stimulus with a natural stimulus (e.g., food) for eliciting such a response, a process that has come to be known as *classical conditioning*. Also during the first half of the twentieth century, psychosomatic medicine gained prominence through the writings of Franz Alexander and Helen Flanders Dunbar as a medical specialty concerned with bodily disorders thought to arise from psychological disturbances.

Thus, the twentieth century arrived as scientists were beginning to understand how the body protected itself and maintained an internal balance through the actions of multiple tissues coordinated by the nervous system in response to both external and internal stimuli, including those originating within the mind. Even then there were some observers such as Hughes (1894) who were eager to put this understanding into practice: "We are approaching an era when the whole patient is to be treated no more only as a part or organ solely. . . . In estimating the causal concomitants and sequences of his diseases, we consider the whole man in his psycho-neuro-physical relations." In more recent times, Engel (1977) has persuasively argued for the need to adopt a biopsychosocial model in health care.

C. Psychoneuroimmunology

The field of psychoneuroimmunology has come into existence in relatively more recent times. The term *psychoneuroimmunology* was first employed by Ader in 1980 to capture what had become growing evidence of the intercommunication between the brain and the immune system. Ader notes that interest in such a link can be traced back to early work by Russian investigators, who, based on Pavlov's work, hypothesized that immune responses could be conditioned and reported preliminary positive findings as early as 1926. Also, the studies of Hans Selye in the 1930s demonstrating that a variety of noxious conditions (stressors) caused endocrine effects and changes in immune tissues were important in the birth of this field. Ader has further traced the emergence of psychoneuroimmunology through the work of Rasmussen (1950s) on the effect of psychological stress on susceptibility to infection, and Solomon's work (1960s) exploring an autoimmune etiology for schizophrenia.

In the 1970s, well-designed conditioning studies, as well as a variety of animal and human studies documenting effects on the immune system by manipulations of the situation or the occurrence of life events affecting the individual's emotional state, further underscored that a link must exist between brain activity and the cells of the immune system. Also beginning in the 1970s, endocrine and ANS connections to cells of the immune system were demonstrated, as was the presence of receptors for neurotransmitters and hormones on immune cells. The early work has been summarized by Ader (1981). In addition, evidence began to accumulate that substances released by cells of the immune systems could affect brain activity and behavior. This, of course, should not be surprising because changes in behavior have always been associated with illnesses. Indeed, the fact that the immune system can affect the brain had already been discovered in the 1930s, when it was observed that vaccination against rabies could cause massive inflammation of the brain. This was because the vaccine was prepared from central nervous system tissue, which led to an immune response against the individual's own brain. Moreover, in the late 1950s and early 1960s, investigators, including Fessel and Hirata-Hibi, Heath, and Solomon, made observations suggesting immune abnormalities in schizophrenic patients, including the possibility that schizophrenia could be an autoimmune disorder. The hypothesis that major psychiatric disorders could result from misdirected immune activity continued to receive attention. Misdirected immune activity has now been implicated in a number of disorders of the brain that affect behavior and mental processes (e.g., childhood-onset obsessive-compulsive disorder). Thus, the picture that has begun to emerge is one wherein life circumstances, by affecting brain activity, can alter immune function, with the potential to have health consequences. Meanwhile, the immune system's response to pathogen stimulation causes the release of substances that can affect brain activity and may even disrupt its functioning.

VII. CONCLUDING COMMENTS

The historical highlights presented in this chapter make it clear that from the outset health has had personal and social consequences. The behavior of the individual, as well as the immediate group, is altered to promote healing. As social conditions change in the history of humanity, so does the health of individuals and the practices of the group with regards to healing. Explanations for illness begin to be elaborated in terms of the deities thought to govern the world. The individual had to be "invaded" by demons for illness to occur and such a thing could be initiated through the ill will of others. Such early ideas already embody the notion of infection and the recognition that social relationships are important in health maintenance. Gradually, as a result of experience with illness and the structure of the body, it became evident that the body was composed of elements of nature that had been transformed and that life had emerged from their dynamic, constantly changing, form of organization. Imbalances could be produced by a multiplicity of factors, and it was apparent that some individuals were more susceptible to being thrown off balance. It seems fair to regard these early ideas as essentially correct at an abstract level, but not as sufficiently grounded in detailed observations about the body or nature in general to permit effective interventions.

Over two millennia, the accumulation of observation and the predominant ideologies that have evolved, along with technological innovations, have permitted a more detailed understanding of how illness occurs. The discovery of microscopic life shifted focus to the level of cells and ultimately to that of molecules. For a time, this level of analysis gained such prominence that it seemed reasonable to ignore other aspects of the individual (e.g., mental life and social circumstances) in the quest to promote health. The pendulum has been gradually and for sometime now moving back toward a more holistic understanding of health, one that is more in line with the views of long ago but that is also well grounded in the complex biology of life. Psychoneuroimmunology represents the most integrative scientific discipline capable of providing an understanding of health in all its complexity and, thus, guiding efforts to promote the health of individuals and communities.

VIII. SOURCES

Ader, R. (ed) (1981), *Psychoneuroimmunology*, New York, Academic Press.

Ader, R. (1995), "Historical perspective on psychoneuroimmunology," in Friedman, H. et al. (eds), *Psychoneuroimmunology, Stress, and Infection*, Orlando, Fl, CRC Press.

Cohen, M.N. (1989), *Health and the Rise of Civilization*, New Haven, Conn, Yale University Press.

Engel, G.F. (1977), "The need for a new medical model: a challenge for biomedicine," *Science*, 196, pp. 129–136.

Fábrega, H. (1997), *Evolution of Sickness and Healing*, Berkley, University of California Press.

Finger, S. (1994), *Origins of Neuroscience: A History of Explorations into Brain Function*, New York, Oxford University Press.

Golub, E.S. (1994), *The Limits of Medicine: How Science Shapes Our Hope for the Cure*, Chicago, University of Chicago Press.

Goodall, J. (1986), *The Chimpanzees of Gombe Patterns of Behavior*, Cambridge, Mass, Harvard University Press.

Hughes, C.H. (1894), "The nervous system in disease and the practice of medicine from a neurologic standpoint," *Journal of the American Medical Association*, 22, pp. 897–908.

Institute of History of Medicine and Medical Research (1973), *Theories and Philosophies of Medicine with Particular Reference to Greco-Arab Medicine, Ayurveda and Traditional Chinese Medicine*, 2nd ed, no. 62, Tughlaqabad, New Delhi.

Kaptchuk, T.J. (1983), *The Web that Has No Weaver: Understanding Chinese Medicine*, New York, Congdon R. Weed.

Keane, R.W. & Hickey, W.F. (eds) (1997), *Immunology of the Nervous System*, New York, Oxford University Press.

Porter, R. (1997), *The Greatest Benefit to Mankind: A Medical History of Humanity*, New York, W.W. Norton & Company.

Sigerist, H.E. (1951), *A History of Medicine*, vol 1, *Primitive and Archaic Medicine*, New York, Oxford University Press.

Sigerist, H.E. (1961), *A History of Medicine*, vol 2, *Early Greek, Hindu, and Persian Medicine*, New York, Oxford University Press.

Solomon, G.F. (2000), *From Psyche to Soma and Back: Tales of Psychosocial Medicine*, Philadelphia, Xlibris Corporation.

Sternberg, E.M. (2000), *The Balance Within: The Science Connecting Health and Emotions*, New York, Freeman.

Val Alphen, J. & Aris, A. (1996), *Oriental Medicine: An Illustrated Guide to the Asian Arts of Healing*, Boston, Shambhala.

Whorton, J.C. (2002), *Nature Cures: The History of Alternative Medicine in America*, New York, Oxford University Press.

CHAPTER 3

Immune System Basics

I. Introduction 23
II. Molecular Self/Nonself
    Discrimination 24
III. Cell Types, Proteins, and Genes 24
IV. Immune System Cell Types
    and Complement 25
V. Immune Cell Functions
    and Inflammation 28
VI. Human Leukocyte Antigen System 31
VII. Antigen Presentation 32
VIII. Natural Killer Cells 34
IX. B Lymphocytes, Antibody Structure,
    and Diversity 35
X. T Lymphocytes 37
    A. T-Cell Receptor Diversity
       and CD Markers 37
    B. Helper T Lymphocytes (CD4$^+$) 38
    C. T-Cell Help of Antibody Production 40
    D. T-Cell Help of Cytotoxicity 41
XI. Cytokines, Chemokines, and Cell
    Adhesion Molecules 42
XII. Lymphoid Organs
    and Leukocyte Traffic 46
XIII. Immune Activation/Deactivation
    and Memory 49
XIV. Development of Immune Functions 50
XV. Measures of Immune Function 51
XVI. Concluding Comments 53
XVII. Sources 55

I. INTRODUCTION

The immune system is a complex set of tissues with mobile elements, whose function is to protect the organism from invasion by exogenous microscopic life forms or particles and to rid the body of defective, damaged, or malignantly transformed cells.

The immune system is composed of a variety of cell types, which form its major organs and circulate through vast networks of lymphatic and blood vessels. The ability of immune cells to circulate, enter extravascular spaces, and recirculate permits their function of detecting pathogens and localizing reactions to them (i.e., causing inflammation). Immune cells coordinate their responses to pathogens by intercommunicating via a wide range of molecules. They also receive molecular signals originating from cells outside the immune system proper.

Immune system responses to pathogens have been classified as either innate (nonspecific) or adaptive (specific). This distinction is largely based on whether the particular immune system component is fully capable of disposing of pathogens from the outset or requires prior exposure to the pathogen and then exhibits a more rapid and effective response upon encountering the same pathogen again. The latter type of response has been further subdivided into responses that are directly mediated by cells (i.e., cell-mediated immunity), or those that are mediated by large circulating molecules (i.e., antibodies) released from some of the immune cells (i.e., humoral immunity).

II. MOLECULAR SELF/NONSELF DISCRIMINATION

The immune system must detect invaders and abnormal cells to accomplish its task of protecting the individual from microscopic invaders or abnormal cells arising within the individual. There must be a mechanism or, more likely, a set of mechanisms for sensing that something that is not a constituent of the healthy organism (defined at the molecular level) is present within the organism. This basic process has been referred to as *self/nonself discrimination*. However, it is important to keep in mind that the healthy human organism shares molecular structures with pathogens and other infectious particles, as well as with cells that are functionally abnormal and potentially capable of harming the individual. Thus, self/nonself discrimination is not a trivial problem and one that can be solved only in relative terms. Perfect discrimination is not attainable. A complementary way in which the immune system appears to become activated is through receptors that sense "danger" in the form of molecular products arising from damaged cells.

III. CELL TYPES, PROTEINS, AND GENES

The various cell types that make up the immune system are classified according to size, shape, appearance, staining properties, and functional properties. The sorts of proteins present on the cell's membrane or within its cytoplasm also serve to differentiate cells into types. At a more fundamental level, the cell types are a reflection of dynamic patterns of gene expression. All cells contain the same genetic material (although in some of the immune cells, sectors of DNA are deleted during the cell's development, in effect causing the rearrangement of the genetic material to form new genes). It is only through the occurrence of different relatively stable patterns of gene activation that cells exhibit different characteristics. In other words, a cell type can be understood as the result of the fact that only some of the genes are turned on, that is, expressing their protein products while the rest are turned off. Only slight differences in the pattern of activation of the estimated 30,000 genes in the human genome give rise to the major cell types composing the immune system.

The switching of genes on and off is the fundamental mechanism whereby stem cells differentiate into the various immune tissue cells, alter their functional state, destroy themselves, or even become transformed into abnormally functioning malignant cells. Genes regulate each other, so activation of one gene can suppress the activity of another through the action of its molecular product within the same cell or on neighboring cells. A major factor regulating gene expression is the chemical composition of the cell's microenvironment, acting through a variety of receptors and intracellular messenger molecules. Thus, the picture that emerges is one of immune cell types as expressions of dynamic patterns of gene activation, which are regulated intrinsically (i.e., the product of one gene within the cell affects the expression of another gene within the same cell) and extrinsically (i.e., products of genes within the cell are affected by the cell's microenvironment, which reflects both other cell gene products and exogenous molecular structures such as foreign antigen).

IV. IMMUNE SYSTEM CELL TYPES AND COMPLEMENT

The major cell types of the immune system derive from hematopoietic stem cells located in the bone marrow. These stem cells have the potential to differentiate into two broad cell types, designated *megakaryocytes* and *leukocytes*. The latter, in turn, can differentiate into polymorphonuclear leukocytes (a.k.a., *granulocytes*) and mononuclear leukocytes. These two broad classes give rise to subtypes: the granulocytes include neutrophils, eosinophils, and basophils, which have been named based on staining characteristics (i.e., colorless, red, and blue, respectively) and differ in functional properties. The mononuclear leukocytes include the monocytes and the lymphocytes. The latter group is composed of the B lymphocytes, T lymphocytes, and the natural killer (NK) cells (large granular lymphocytes). Other cell types thought to arise from myeloid and lymphoid precursors include the following: the mast cells, which are tissue bound and thought to be

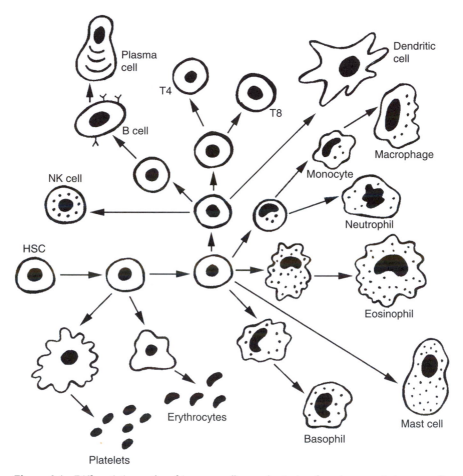

Figure 3.1 Differentiation paths of immune cell types beginning from hematopoietic stem cells (HSC) in bone marrow. NK, natural killer.

related to the basophil subpopulation; the macrophages (tissue-bound monocytes); and the dendritic cells, which have limited mobility and appear to arise from a variety of precursors. Figure 3.1 illustrates the differentiation trajectories, which, as previously discussed, reflect changes in gene-activation patterns within the cells.

The immune response repertoire also includes the *complement* system, which consists of some 20 proteins that are produced primarily in the liver and released into the circulation (Figure 3.2). The concentration of some of these proteins in the blood is increased during the acute response to infection. Some of these proteins are capable of reacting with molecular structures unique to the cell wall of bacteria or

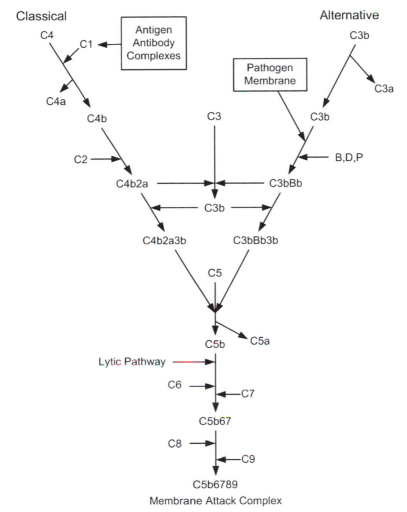

Figure 3.2 Classical and alternative complement activation pathways. Complement components (C1 through C9) interact with each other in a cascade fashion that leads to the formation of proinflammatory and cytolytic elements. B, factor B; D, factor D; P, properdin.

with antibody already attached to the bacterial membrane. The complement system facilitates ingestion of pathogens by phagocytes and initiates a process whereby the pathogen's cell wall is punctured and destroyed. The most pivotal complement component is C3, present in plasma at concentrations of around 1–2 mg/ml. It undergoes slow spontaneous cleavage to yield C3b, which can be quickly inactivated unless it interacts with factor B and factor D to form C3bBb (C3 convertase), which can further promote the cleavage of C3. Microorganisms can promote this process with the result that a large amount of C3b forms on the microbial surface and serves to activate C5 convertase (C3bBb3B). The latter in turn cleaves C5 to form C5b, which along with C6, C7, C8, and C9 cause the formation of a membrane attack complex and brings about cell lysis. This pathway of activation is known as the *alternative pathway*. There is also a *classical* pathway in which antibody–antigen complex interacts with C1 and jointly catalyzes the cleavage of C4. The cleavage products, in conjunction with C2, lead to the formation of C4b2a, another enzyme capable of splitting C3 and initiating the cascade brought on by the alternative pathways. It should be noted that complement activation and phagocytosis, so-called *nonspecific immune responses*, can be targeted and enhanced by specific immune responses such as the production of antibodies.

V. IMMUNE CELL FUNCTIONS AND INFLAMMATION

Neutrophils are granular leukocytes (i.e., granulocytes) possessing a nucleus with three or more lobes (i.e., polymorphonuclear). They are short-lived (hours to days) cells. It has been estimated that about 60% of hematopoietic activity in the bone marrow is directed toward neutrophil production. Neutrophils are involved in the ingestion (phagocytosis) of pathogens such as bacteria (Figure 3.3). They are capable of phagocytosis on their own (i.e., without the participation of any other immune mechanism) or by relying on antibodies or the complement components to recognize and ingest pathogens. This is accomplished via receptors on the neutrophil for a part of the antibody molecule and for complement molecules.

Eosinophils are cells that contain granules and irregularly shaped nuclei. The granules contain proteins that are cytotoxic and are particularly effective in ridding the organism of the class of pathogens known as *parasites*.

Basophils have granules and irregular nuclei. They circulate and may be the precursor cell to the tissue-bound *mast cells*, which participate in local inflammation and allergic reactions. The granules contain a variety of substances (e.g., histamine, serotonin, heparin, neutral proteases and hydrolases, platelet-activating factor, prostaglandins, and leukotrienes) that alter vascular permeability in the region and allow leukocytes to enter tissues and mount local immune responses. Mast cells are numerous at sites such as the skin, respiratory tract mucosa, and gastro-intestinal tract, where pathogens would be first encountered as they penetrate the organism. Mast cell degranulation (Figure 3.4) can be triggered by antibodies, particularly a class of antibody designated *immunoglobulin E* (IgE), again, an instance of a nonspecific response (mast cell degranulation) activated specifically by anti-bodies. Degranulation can also be triggered by products of the complement system or by the release of peptides from nerve terminals in the region, a clear example of an immune response initiated by a neural signal.

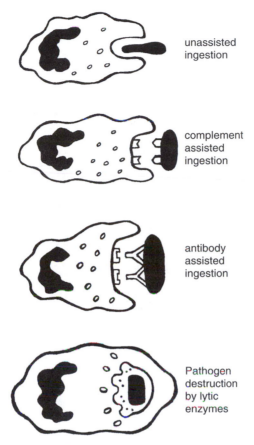

unassisted
ingestion

complement
assisted
ingestion

antibody
assisted
ingestion

Pathogen
destruction
by lytic
enzymes

Figure 3.3   Neutrophil ingestion of pathogen, a process that is aided by complement components and antibody.

The neutrophils, eosinophils, and basophils are key players in the inflammatory process. This is the process whereby any type of tissue damage can cause the release of proinflammatory mediators that cause vasodilation (the basis of redness and warmth), increased vascular permeability, and movement of leukocytes (particularly neutrophils) into the tissue (the basis of swelling). Multiple mediators participate in the process of inflammation: the plasma proteases (i.e., complement, kinins, and fibrinogen), lipid mediators (prostaglandins, leukotrienes [both products of arachidonic acid metabolism], and platelet activating factor), peptides (substance P, vasoactive intestinal peptide, somatostatin, and calcitonin gene–related peptide), amines (histamine and serotonin), and nitric oxide. Cytokines are also involved including the following: interleukin-1 (IL-1), IL-4, IL-6, IL-8, tumor necrosis

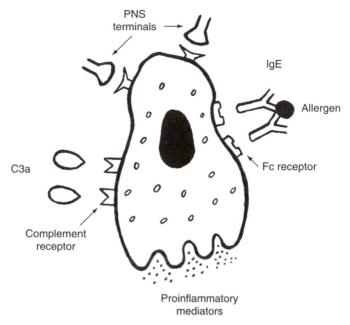

Figure 3.4 Mast-cell degranulation triggered by antigen–antibody (immunoglobulin E [IgE]) complexes, complement cleavage products (C3a), and neuropeptides released from peripheral nervous system (PNS) terminals.

factor (TNF), interferon-γ (IFN-γ), and IL-12. This wide array of molecular signals acts to attract neutrophils to the damaged area. If the source of irritation is not successfully eliminated, acute inflammation can turn into a chronic inflammation and the tissue can be further infiltrated by monocytes and lymphocytes.

Monocytes have nuclei with a more regular appearance that gives the impression of being folded. They also contain granules, which can destroy bacteria after they are ingested by the monocyte. Monocytes can become tissue bound and enlarged. These cells are then referred to as *macrophages* (large eaters) to highlight enhanced phagocytic capability. Macrophages belong to a select group of cell types that are known as *antigen-presenting cells* (APCs). This group also contains the *dendritic cells* and *B lymphocytes*, which are also the antibody producing cells. APCs are abundantly evident at portals of entry into the body (e.g., skin and mucosa). *Dendritic cells*, in particular, are extensively distributed. They go by different names depending on where they are found. In the skin and mucosal surfaces, they are known as *Langerhans' cells*. In the heart, kidney, intestines, and lung, they are referred to as *interstitial dendritic cells*, and *interdigitating cells* if found in the thymus. Dendritic cells that have migrated out of their original site and move toward a lymph node in the afferent lymphatics are called *veiled cells*.

Lymphocytes constitute a subpopulation of leukocytes, which include the NK cells, the T lymphocytes, and the B lymphocytes. The latter two have the remarkable ability to bind antigen in a relatively specific manner. Subsets of the B and T lymphocytes are capable of binding (i.e., recognizing) only a narrow range of short amino acid sequences (epitopes) out of the trillions that are possible. This is accomplished through membrane-bound receptors, which possess a region with highly variable structure known as the *antigen-binding site*. In the case of the B lymphocytes, the "receptor" is sensitive to shape (i.e., conformation) of large molecules, and it is actually the antibody molecule that the cell is initially programmed to produce. In the case of the T lymphocyte, there is an actual receptor molecule whose function is solely to bind specific antigens, appearing in the form of short amino acid sequences located within a groove of the Human Leukocyte Antigen (HLA) molecules.

VI. HUMAN LEUKOCYTE ANTIGEN SYSTEM

There are six genes located on human chromosome 6, which are referred to as the HLA system (Figure 3.5), or in other species as the major histocompatibility complex (MHC). When expressed on the surface of cells, this set of genes makes them recognizable as self-tissues. These are the molecular structures that are examined when organ donors are matched to recipients to minimize rejection of the transplanted organ by the recipient's immune system. There are two classes of HLA molecules, each consisting of up to six distinct molecules. HLA class I molecules are coded from loci A, B, and C in chromosome 6. HLA class II molecules are coded from loci DR, DP, and DQ, also located on chromosome 6. These locations are said to be polymorphic, that is, there are many versions (i.e., alleles) of these genes. For each location, the gene from the mother may differ from that of the father and both are expressed. Thus, individuals differ not only with respect to specific alleles, but also with respect to whether they are homozygous or heterozygous (same allele from both parents or different ones). The number of distinct HLA molecules expressed on the cells of the individual can range from 3 to 12.

The HLA molecules are also referred to as *antigens*. The word *antigen*, which is a condensed version of "*anti*body *gen*erator," is unfortunately used to refer to molecular structures from both pathogens and malignant cells, as well as those from healthy self-tissue. This can cause confusion if one tends to equate *antigen* with *pathogen*. Essentially, antigen refers to any macromolecular structure irrespective of its source. However, in the interest of clarity, the term *antigen* will only be used to denote molecular structures originating in pathogens or abnormal self-tissue, and *self-antigen* will refer to molecular structures that originate in healthy self-tissue. The class I molecules are displayed on the cell surface of all nucleated cells in the body. The class II molecules are primarily displayed on the cell membrane of APCs.

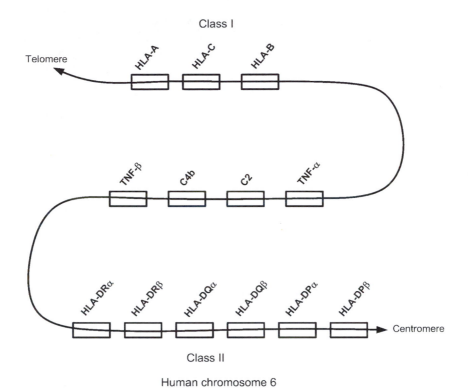

Figure 3.5   Relative location of genes coding for class I and class II Human Leukocyte Antigen (HLA) proteins.

VII. ANTIGEN PRESENTATION

An understanding of how antigen is presented makes it possible to appreciate how the immune system detects pathogens or abnormal cells and then mounts specific responses to them. HLA molecules contain a groove or small area that can be occupied by a protein fragment (i.e., peptide) consisting of a sequence of 10–14 amino acids. This groove must be occupied before the HLA molecule will move from the cytoplasm to the cell membrane (Figure 3.6). Most of the fragments displayed by HLA molecules are self-antigens in that they originate from proteins that are required for the healthy function of the cells.

Fragments can also originate from proteins that are either part of a virus that has infected the cell or the results of an abnormal change to the cell (malignant transformation). Class I HLA molecules will display such fragments on cell

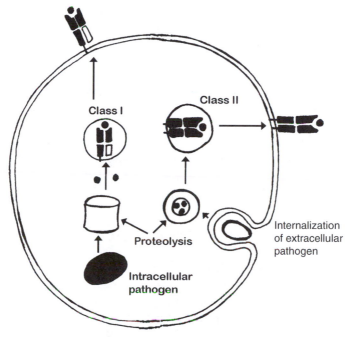

Figure 3.6 Processing of pathogen-derived proteins to create fragments that are then displayed on Human Leukocyte Antigen molecules at the cell's surface.

membranes and thus create part of the conditions necessary for the recognition that the cell is somehow compromised. In addition, the cells known as *APCs* are capable of ingesting microorganisms that are extracellular or their products (e.g., toxins). Proteins derived from such sources give rise to fragments that are held in the groove of HLA class II molecules and displayed on the membrane of the APCs. Antigen presentation occurs primarily in secondary lymphoid tissue (e.g., spleen, lymph nodes, and mucosa-associated lymphoid tissue [MALT]). B lymphocytes are unique in that they both present and recognize antigen. The primary target of such presentation is the subpopulation of lymphocytes known as *T lymphocytes.*

Thus, what emerges is a situation in which class I molecules display peptides from healthy self or intracellular pathogens on the surface of most cells of the body. In addition, class II molecules display peptides originating from healthy self-tissues or extracellular pathogens on the surface of a subset of cells known as *APCs.* This amounts to a process whereby there is constant appraisal of whether the protein composition of the individual has shifted significantly from what defines health. As long as HLA molecules are displaying self-antigens, relatively speaking, everything is all right.

Before proceeding to further examine how this process of recognition is ultimately accomplished (and results in an immune response), it may be of value to comment further on this stage in the process of discrimination between self-antigens and other antigens, in other words, between fragments of proteins that define healthy tissue versus those that do not, given that all proteins consist of a series of amino acids irrespective of their source. The specific order and length of the sequence of amino acids give rise to a protein's structure and ultimately define its function. Each individual is composed of a finite set from the possible universe of proteins, and only those proteins need to be recognized as self. It is not necessary to inspect the entire protein molecule, sequences (peptides) of 10–14 amino acids are sufficient to allow the discrimination process to work well enough to detect deviations from self-antigen.

VIII. NATURAL KILLER CELLS

NK cells are large granular lymphocytes capable of cytotoxicity. They participate in the process of eliminating cells that have been infected or somehow become defective. NK cells bind to the target cells using receptors (NK-RP1) for carbohydrate structures on target cell membranes (Figure 3.7). They also have a receptor for part of the HLA class I molecule and bind to it when self-antigen is present in the groove of the HLA class I molecule. Binding to the HLA class I molecule prevents the NK cell from releasing lytic enzymes that damage the target cell's membrane and result in cell death. Essentially, NK cells read the absence of self-antigen or of HLA class I molecules as a signal for releasing lytic enzymes. In addition, NK cells have receptors for a part of the antibody molecule, so they will destroy cells that are covered with antibody, a process known as *antibody-dependent cellular cytotoxicity* (ADCC).

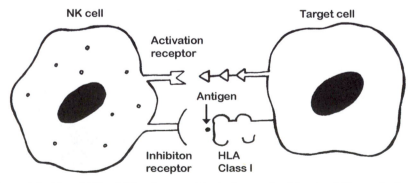

Figure 3.7 Natural killer (NK) cell cytotoxicity is downregulated by class I Human Leukocyte Antigen expression on target cells.

IX. B LYMPHOCYTES, ANTIBODY STRUCTURE, AND DIVERSITY

B lymphocytes are the cells that produce antibodies. They undergo much of their maturation within the bone marrow and complete it in the lymph nodes within the microenvironment of the lymphoid follicle. In the bone marrow, the progenitor B cell already shows evidence that the genes coding for the variable region of the antibody molecules are rearranging.

Antibodies belong to one of the five protein groups found in the liquid part of the blood. They belong to the γ-globulin group and they are referred to as *immunoglobulins* (Ig). The basic structure of antibody (Figure 3.8) consists of four separate protein chains; two are longer (heavy chains), and two are shorter (light chains). The arrangement of these proteins is such that the heavy chains are adjacent and bound by disulfide bonds. The light chains are in turn bound to the heavy chains beginning at the variable region end of the heavy chain. The light chain also possesses a variable region. The groove between the variable regions (L and H) on all arms of the antibody molecule serves as the antigen–binding site. The part of the antibody molecule containing both chains (L and H) is known as the *Fab region*; the part containing only the heavy chains is known as the *Fc region*.

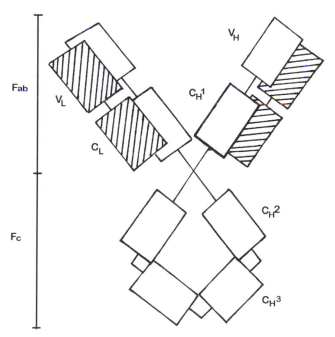

Figure 3.8   Structure of antibody heavy (clear) and light (striped) chains. Antigen–binding site is located between the variable region of the light (V_L) and heavy (V_H) chains. See text for additional details.

A hinge or flexible region occurs between the Fab and Fc regions, allowing the antibody molecule to bend and look roughly like the letter *Y*. The variable regions give the antibody molecules their specificity. The constant regions serve to form the class to which the antibody belongs. There are five main classes of antibody molecules: IgG, IgA, IgM, IgD, and IgE. Four subclasses of IgG have been recognized, two each of IgA and IgM. Differences in the amino acid sequences of the heavy chains serve to define the subclasses (or isotypes).

As noted earlier in this chapter, one of the earliest events in the maturation of B lymphocytes is the rearrangement of the genes coding for the variable regions of the heavy chains. A similar process occurs in the case of the variable region of the light chain. Genes are assembled by randomly deleting regions of the DNA and recombining specific segments of the DNA (Figure 3.9). In the case of the heavy chains, a variable segment (V), a diversity segment (D), and a joining segment (J) are randomly selected and brought together with the constant segment (C) to form a specific gene. For the light chain gene, two segments, a variable (V) and a joining (J) segment, are randomly selected for recombination with the constant segment. Essentially, the random assembly of DNA segments within the cell to form a functioning gene, which is then expressed, is what underlies the diversity of antibody molecules. Some degree of imprecision in the joining of segments, as well as RNA editing, adds further diversity (i.e., sequence heterogeneity) to the variable regions. In humans, the number of unique combining sites is estimated to be on the order of about 10^{12}. B cells, which tightly bind to self-antigen, undergo elimination in the bone marrow. Nonetheless, antibodies with detectable binding to self-antigen escape the deletion process in the bone marrow and must somehow be regulated in the periphery to prevent damage to healthy tissues.

It is important to underscore that genetic rearrangement could be lethal if it happened at random to any segment of DNA. It could result in change to a

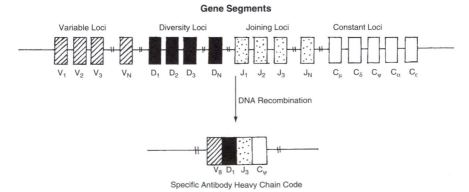

Figure 3.9   Rearrangement of DNA segments in B cells gives rise to unique antigen–binding sites. An analogous process occurs in T cells and leads to the formation of specific T-cell receptors.

life-sustaining enzyme so that it would no longer function. Thus, such rearrangement must be highly regulated and restricted. It appears restricted by cell type, the cell's developmental stage, and the phase of the cell's cycle. Moreover, even though gene rearrangement in both B and T cells relies on a common enzyme system (recombinase machinery), it is restricted to the antibody variable region genes in the B cells and the T-cell receptor (TCR) variable region genes in T cells.

X. T LYMPHOCYTES

Lymphocytes that are "destined" to become T lymphocytes are thought to leave the bone marrow with a surface molecule that facilitates adhering to the arterial and venous capillaries in the thymus. T lymphocytes are so designated because they become functional in the thymus. Epithelial cells of the thymus produce hormones that promote T-cell maturation. During this process, the TCR is assembled and expressed. Specific T-cell surface marker molecules (e.g., CD3, CD4, and CD8) are expressed. "CD" signifies "cluster of differentiation," which along with an identifying number is used to denote specific cell surface molecules according to the International Leukocyte Workshops.

TCRs bind antigen that appears in the groove of the HLA class I or class II molecules. Therefore, the receptor must "fit" the HLA molecules before it can bind antigen. If such binding does not occur in the thymus, the cell dies, through the process known as *apoptosis*. Similarly, the cell is eliminated if it binds tightly to self-antigen (all antigen encountered in the thymus environment is "assumed" to be self-antigen). T-cell elimination in the thymus is estimated to be close to 95%. Only a small minority of the T-cell population is deemed capable of protecting the organism against pathogen. It is worth noting that the body appears to be engaging in activity that could be construed as wasteful (i.e., produce and kill 95% of the T lymphocytes) to select a subset that will be adequately poised to recognize antigen in a given individual. This process of elimination is necessary for the development of "tolerance," a term used to describe the fact that normally T cells do not mount full-blown responses to self-antigens. Nonetheless, T cells, which are capable of reacting to self-antigen, can be found in the circulation. Therefore, additional mechanisms of tolerance induction have been proposed.

A. T-CELL RECEPTOR DIVERSITY AND CD MARKERS

The diversity in TCR structure that allows relatively specific binding reflects the fact that a process of random gene rearrangement occurs within each T cell, as was the case for the B cells. The TCR is a heterodimer (i.e., composed of two polypeptides, predominantly α and β) selected from four possible polypeptides (α, β, γ, and δ) available in the DNA (Figure 3.10). The TCR is associated with a set

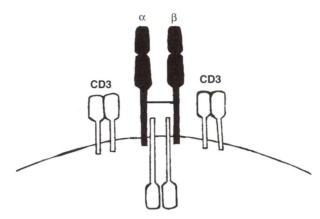

Figure 3.10   Basic T-cell receptor structure.

of transmembrane proteins collectively designated the *CD3 complex*. The variable region (the antigen–binding site) of the TCR is assembled from random rearrangements of genetic material to form different receptor genes within each T cell in a manner analogous to how the antibody molecules are generated within the B cells. The β chain is coded from a gene resulting from the recombination of three distinct segments of DNA (a variable region, a diversity region, and a joining region) randomly selected from numerous possible segments and then combined with a constant segment. The α chain is coded from two segments (a variable region and a joining region), also randomly selected from numerous possibilities, which undergo recombination and then combine with a constant segment. The diversity of structure possible from these rearrangements is on the order of 10^{16}. In addition to the process of TCR expression, maturation in the thymus also results in expression of cell surface molecules that are central to the function of T cells. T cells that express the CD4 molecule (CD4$^+$) are known as *helper T cells* and play an essential role in immune function. They are the cells that are most damaged by human immunodeficiency virus (HIV) infection, which can lead to the collapse of immune function. The other major group of T cells expresses the CD8 molecule (CD8$^+$). The CD8$^+$ subset is capable of cytotoxicity when activated by APCs or CD4$^+$ cells. CD4$^+$ T cells recognize antigen presented in HLA class II molecules, whereas CD8$^+$ T cells recognize antigen presented in HLA class I molecules. Once activated, CD8$^+$ cells are able to destroy any cell of the body that appears infected or somehow altered from the healthy state.

B. HELPER T LYMPHOCYTES (CD4$^+$)

After T cells complete development in the thymus, they emerge as two major subclasses recognizable based on cell surface marker molecules. As noted earlier in this chapter, these are the CD4$^+$ and CD8$^+$ T lymphocytes, also referred to as *helper* and *cytotoxic T lymphocytes*, respectively. The helper T cells remain in a "naive" state until they encounter antigen and appropriate costimulation. They are then transformed into effector/memory cells. This transformation essentially alters how the cell continues to respond to costimulation from other cells in the microenvironment, how rapidly it proliferates, and how easily it moves in and out of tissues. However, if the "naive" cell binds strongly to antigen without costimulation, it reacts by becoming "anergic" (i.e., entering a state of relative deactivation), which appears to be an additional mechanism for the development of tolerance.

The CD4$^+$ population appears to have at least two functional subpopulations, the Th1 and the Th2 cell types produced from a precursor designated Th0 (Figure 3.11). The relative proportion of these cells, or the functional balance

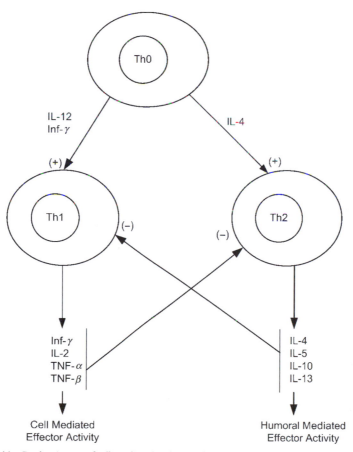

Figure 3.11 Predominance of cell-mediated or humoral-mediated activity is regulated by the balance between Th1 and Th2 cytokine profiles. IFN, interferon; TNF, tumor necrosis factor.

between them, is also affected by the molecular composition of the cell's micro-environment. They are differentially enhanced by exposure to particular cytokines and peptides. They tend to antagonize each other, so Th1 suppresses Th2 and vice versa. This antagonism may be the source of the T-cell suppression that has been observed. Th1 and Th2 cell types produce different effects with respect to B-lymphocyte activation. Th1 stimulates B lymphocytes to secrete primarily IgGs, whereas Th2 promotes secretion of IgA, IgE, and IgG4. Predominance of Th1 favors cell-mediated effector activity, whereas that of Th2 favors humoral effector activity.

C. T-CELL HELP OF ANTIBODY PRODUCTION

T-lymphocyte helper function is central to all facets of immune responsiveness including nonspecific responses. B cells are capable of being activated by membrane constituents of pathogens independent of T-cell participation. Such B cells are often found in intestinal mucosa and in the marginal zone of the spleen. Essentially, they act more like cells that are part of innate immunity. However, the scope of such a response cannot adequately protect the organism. T-cell regulation of B-cell activation appears essential to effective function of humoral immunity. The process of T-cell regulation of the production of specific antibody goes as follows: APCs (e.g., dendritic cells) ingest and lyse proteins from pathogens. They move near the region of lymph nodes where naive $CD4^+$ T lymphocytes reside. APCs present antigen to the T cells in the region. Those that bind sufficiently to the antigen and receive costimulation become activated, begin proliferating, and release cytokines that can facilitate B-cell activation and specific antibody production. However, this outcome depends on B cells in the region having bound the same antigen as the T cell, internalized it, and then displayed it on its HLA class II molecules (Figure 3.12). The activated helper T cells can now recognize the antigen on the B cell and further bind to other receptors on the B cell. This coupling of T and B cells recognizing the same antigen enables the cytokine-induced activation of the specific B cell and prevents activation of other nearby B cells that are not specific for the antigen that has been detected. This B-cell activation results in proliferation and differentiation into memory cells and antibody-producing cells (plasma cells). It also triggers a process whereby the B cell improves its antigen-binding capability by further mutating the genes coding for the variable region in an iterative manner, leading to better binding of the specific antigen. During the signaling, which causes B-cell activation, the profile of cytokines released can cause heavy-chain class switching, that is, cause the constant region of the heavy chain to change from IgM isotype to either IgA, IgG, or IgE.

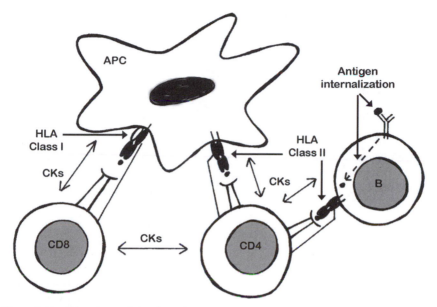

Figure 3.12 T-lymphocyte help of specific antibody production and cytotoxicity. See text for details. APC, antigen-presenting cell; HLA, Human Leukocyte Antigen; CK, cytokine.

D. T-Cell Help of Cytotoxicity

Activated CD4$^+$ T lymphocytes can migrate out of the lymph nodes and look for APCs in the tissues that are displaying its specific antigen. If antigen is found, the CD4$^+$ cell, through the release of various cytokines, will initiate chemotaxis of other cells toward the area that are capable of causing inflammation and ingesting the pathogen (i.e., it will attract neutrophils and monocytes). This process is known as the *delayed hypersensitivity reaction*. It is a good example of T-cell enhancement of nonspecific immune responses.

Activated CD4$^+$ cells are also capable of promoting cytotoxicity by activating CD8$^+$ T cells or enhancing NK activity. The latter, another example of a nonspecific response (i.e., antigen independent), requires only cytokine stimulation of the NK cells by the CD4$^+$ cells. In contrast, CD8$^+$ activation entails binding of the CD8$^+$ TCR to specific antigen displayed on the HLA class I molecule of the same APC (or one that is nearby), where the CD4$^+$ cell is binding to antigen within the HLA class II molecule (see Figure 3.12). Cytokine release (e.g., IL-2) from the CD4$^+$ cells serves to activate the CD8$^+$ cell. Once the CD8$^+$ T cell is activated, it becomes capable of specific cytotoxicity (i.e., the killing of cells infected or displaying the particular antigen) independent of the presence of CD4$^+$ T cells in the vicinity.

XI. CYTOKINES, CHEMOKINES, AND CELL ADHESION MOLECULES

In the preceding section, reference was made to the fact that CD4$^+$ cells change their functional characteristics when exposed to antigen and costimulating signals. CD4$^+$ cells are induced to proliferate (clonal expansion), change their migratory patterns, and become capable of influencing other cells (e.g., B cells). Costimulation is accomplished via the release of cytokines (communication molecules) and the expression of receptors capable of reading their messages (i.e., responding to their presence in the cell's microenvironment).

Other cell surface molecules promote cell-to-cell interactions (as between a T cell and an APC) and regulate the homing of leukocytes to sites of infection (Figure 3.13). More than 160 cell surface molecules have been recognized on lymphocytes. They serve as channels for receiving additional regulatory signals (costimulation) during the activation process. To date, more than 20 such molecules have been shown to augment the T-cell proliferative response to antigen. Molecules that play an inhibitory role have been recognized as well. When both activating and inhibiting receptors are simultaneously engaged by their respective ligands, the net outcome is determined by the relative strength of the opposing signals.

Cytokines are proteins secreted by cells that can act on the cell itself (autocrine) or on other nearby cells (paracrine). They bind to specific receptors. There are numerous cytokines (Table 3.1), but some are better characterized than

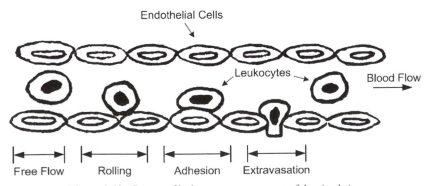

Figure 3.13   Process of leukocyte movement out of the circulation.

Table 3.1

Cytokines

Name	Major sources	Selected actions
Interleukins		
IL-1	Monocytes Macrophages	Promote inflammation

IL-2	Helper T cells (CD4$^+$)	T-cell growth
		Enhance lymphocyte
		  responsiveness (CD4$^+$, NK,
		  and B cells)

IL-3	Helper T cells (CD4$^+$)	Colony-stimulating factor Proliferation of hematopoietic cells
IL-4	Helper T cells (CD4$^+$,Th2) Mast cells	B-cell growth Switch to IgE production Inhibit cell-mediated immunity
IL-5	Helper T cells (CD4$^+$) Mast cells	B-cell growth Switch to IgA production Eosinophil differentiation

IL-6	Vascular endothelial cells;
	  monocytes/macrophages/
	  fibroblasts
	Activated T cells
		Promote inflammation
		Induce acute-phase proteins
		Promote hematopoiesis
		Promote antibody synthesis
		Activate helper T cells

IL-7	Stromal cells of bone marrow and
	  thymus
		Stem cell differentiation into
		  pre-T lymphocytes and
		  B lymphocytes

IL-8	Macrophages Endothelial cells Lymphocytes	Act as a neutrophil chemoattractant
IL-9	Activated helper T lymphocytes	Selective enhancement of helper T-cell clones

IL-10	T cells
	Monocytes/macrophages
	Activated B cells
		Cytokine synthesis
		Inhibitory factor for Th1 cell-
		  mediated activity
		Decrease HLA class II molecule
		  expression

IL-11	Stromal cells in bone marrow	Promote megakaryocyte growth
IL-12	T cells, macrophages Dendritic cells	Promote cytotoxic activity by CD8$^+$ and NK cells
IL-13	Activated T cells	Inhibit proinflammatory cytokine production Enhance IFN-γ production Activate B cells

(Continues)

Table 3.1 (*Continued*)

IL-14	Dendritic cells in follicles T cells in germinal centers of lymph nodes	B-cell proliferation Enhanced B-cell memory
IL-15	Monocytes and various other cells	Mast cell growth T-cell growth Promote NK activity
IL-16	Monocytes Helper T cells (CD4$^+$) Mast cells Eosinophils	Chemoattractant for CD4$^+$, monocytes, and eosinophils

IL-17	Activated memory CD4$^+$ cells	Induce synthesis of cytokines that
		  promote T-cell-dependent
		  inflammation

IL-18	Macrophages	Promote inflammation enhances
		  Th1 response enhances NK
		  activity

Interferons		
IFN-α	Macrophages B cells	Prevent viral replication Induce NK activity
IFN-β	Fibroblasts	Prevent viral replication
IFN-γ	Activated T cells and NK cells	Activate macrophage phagocytosis Promote Th1 activity
Tumor necrosis factors		
TNF-α (cachectin)	Monocytes Macrophages T cells B cells NK cells	Promote inflammation Facilitate leukocyte recruitment Lysis of some tumor cells Cause wasting syndrome
TNF-β (lymphotoxin)	Activated lymphocytes	Facilitate cytotoxicity
Transforming growth factor		
TGF-β1, 2, 3	T cells Macrophages Other cells	Inhibits activation Anti-inflammatory
Colony-stimulating factors		

Granulocyte-
  macrophage CSF
MEG-CSF
	T cells
	Fibroblasts
	Endothelial cells
	Macrophages
		Growth factor for granulocytes/
		  monocytes
		Megakaryocytes

EO-CSF M-CSF		Eosinophils Macrophages
Other factors		

Migration inhibition
  factor
	Lymphocytes	Inhibit migration of monocytes
		  and macrophages, and induce
		  proinflammatory cytokines

(*Continues*)

Table 3.1 (*Continued*)

Leukocyte inhibitory factor	Activated T cells	Prevent neutrophil, eosinophil, and basophil migration
Macrophage inflammatory proteins-1 (MIP-1α) (MIP-1β)	Macrophages Monocytes Lymphocytes Fibroblasts	Promote fever Stimulate phagocytes Act as leukocyte chemoattractant

Note: HLA, Human Leukocyte Antigen; NK, Natural Killer

others. Many are referred to as *interleukins* (IL). The name derives from the initial observation that they were produced by leukocytes and were used for intercommunication among leukocytes. However, it is now known that such cytokines can be produced by other cells as well. The cytokine family includes molecules such as IL-1, IL-2, IL-3, IL-4, IL-5, IL-6, and TNF-α to name a few. Interferons are also included in this group because they share a number of features with cytokines. Interferons (IFN-α, IFN-β, and IFN-γ) gain their name from the early observation that they interfered with viral replication.

Cytokines exhibit the phenomenon of pleiotropy (one cytokine can exert many different effects depending on target cell type) and redundancy (many different cytokines can induce similar effects). For instance, IL-2 can induce T-cell growth, B-cell immunoglobulin synthesis, and NK-cell activation. T-cell growth can also be induced by IL-4, IL-7, IL-9, and IL-15. To complicate matters further, there are soluble receptors for cytokines, which can bind the molecules ("intercept them") and prevent their interaction with the membrane-bound receptor, thus, blocking their intracellular effects. In the case of IL-1, there is even an endogenous receptor antagonist (IL-1RA), which essentially blocks the actions of IL-1α or IL-1β at the membrane bound receptor.

As noted earlier in this chapter, helper T cells can be differentiated into two major types (Th1 and Th2) with distinct cytokine production profiles. In humans, Th1 cells produce IFN-γ more exclusively, whereas Th2 cells produce IL-4, IL-5, and IL-9 more exclusively. Both types of cells are capable of producing other cytokines (e.g., IL-3). Th1 cells are more intimately involved in cell-mediated immunity (inflammatory responses, delayed type hypersensitivity, and cytotoxicity), and Th2 cells in humoral immunity (antibody production).

Of all the cytokines, TNF-α and IL-1 exhibit the broadest spectrum of effects thus far documented. They are considered proinflammatory cytokines. TNF-α appears to promote normal lymphoid tissue development and is a key cytokine in resistance to intracellular pathogens. TNF-α is directly cytotoxic to some tumor cells but has also been observed to promote tumor growth. Moreover, if overproduced, TNF-α can be quite damaging and it has been implicated in the pathogenesis of septic shock.

IL-1 can increase the production of other cytokines (IL-6, TNF-α), promote T-cell development and B-cell maturation and differentiation, and participate in leukocyte recruitment to infected sites. In addition, IL-1 can have systemic effects, such as producing fever, inducing acute-phase proteins, and activating the hypothalamus–pituitary–adrenal axis. Here one encounters important effects of immune activation on other organ systems, most notably the brain.

Another group of cytokines are known as *chemoattractant cytokines* (or *chemokines*), which induce directional migration and activation of leukocytes and, thus, play a major role in inflammatory responses. They are produced not only by leukocytes but also by other cell types. They help to direct and retain cells in a region where pathogen is present. For instance, migration inhibition factor (MIF) restricts leukocyte migration out of an infected area.

Cell adhesion molecules (CAMs) provide leukocytes and other cells with the ability to attach to each other. Cells can then remain in place, move out of the circulation into surrounding tissues (extravasation), or interact in proximity, as during antigen presentation or cytotoxicity. There are three major groups of CAMs: the selectins, the integrins, and the immunoglobulin-supergene family. They are glycoproteins and include molecules such as the following: endothelial–leukocyte adhesion molecule-1 (E-LAM-1), which is a selectin involved in slowing the movement of leukocytes; leukocyte function–associated antigen-1 (LAF-1), an integrin that allows the attachment of T cells to other cells; and intercellular adhesion molecule-1 (ICAM-1), a member of the Ig supergene family, which is expressed on endothelial cells and acts as a receptor for selectins and integrins expressed on leukocytes.

Evidently, the intercommunication occurring in the process of mounting an immune response involves rather extensive signaling via a wide array of molecules acting on multiple receptors before, during, and after the encounter with specific antigen. The utter complexity of this process is what enables effective self/nonself discrimination, that is, tolerance to self-antigen and vigorous well-targeted and transient effector activity against foreign antigen.

XII. LYMPHOID ORGANS AND LEUKOCYTE TRAFFIC

The immune system must maintain surveillance over the entire organism, because pathogens can gain entry at any point over the entire surface of the organism. Moreover, given that lymphocytes have unique antigen-recognition capabilities, it is not possible to simply station cells. Cells must move about in search of specific antigen. Immune cells originate in the primary lymphoid organs, which include the bone marrow and the thymus. In humans, after early fetal life B lymphocytes differentiate in the bone marrow, whereas T lymphocytes undergo the final stages of differentiation in the thymus. The immune cells then begin to

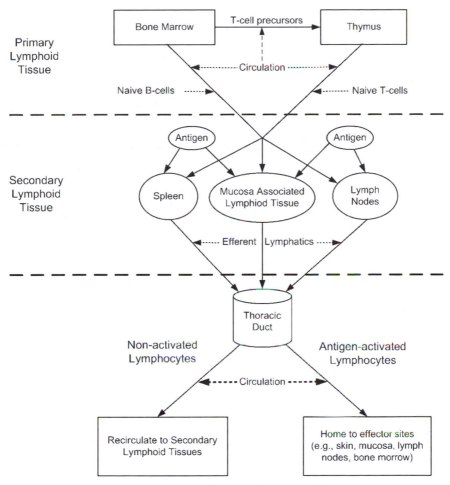

Figure 3.14 Overall pattern of leukocyte traffic.

circulate throughout the body (Figure 3.14). Immune cells are able to leave the circulation and enter tissues. They then return to the blood by first entering the afferent lymphatics and passing through the lymph nodes into the efferent lymphatics, and then they move toward the thoracic duct where they reenter the circulation through the left subclavian vein. As cells are channeled through these vessels, they can attach to molecular structures (adhesion molecules) expressed on high endothelial venule cells, roll along the inner surface of the vessel, be held in place, or be directed into the extravascular space. It is noteworthy that naive lymphocytes cannot migrate into tissue. They must first be activated by encountering antigen in the spleen, lymph nodes, or MALT.

The lymph nodes are major hubs of activity. The number of "naive" lymphocytes that enter a single lymph node from the circulation is estimated at 14,000 per second. This does not include the number recirculating through the lymphatic vessels. The number of lymphocytes in the circulation at any moment is estimated to be on the order of about 10^{10}. Lymph nodes are supplied by a single artery. High endothelial venule cells line the finer branching of the node's vascular supply and allow the passage of naive B and T lymphocytes into the node. The lymph nodes are composed of a B–cell area (cortex) where they cluster into follicles, a T–cell area (paracortex), and a medulla, which contains B, T, and plasma cells. Antigen presentation and lymphocyte activation occurs in the lymph nodes, as previously described. Movement out of the lymph nodes is temporarily halted for cells that recognize antigen.

The spleen is the largest aggregation of lymphoid tissue in the body and is another key structure in which antigen present in the blood is processed. The spleen receives a blood supply and has efferent lymphatic vessels, but no afferent ones. Arterial branches are covered with lymphocytes (i.e., the periarteriolar lymphocyte sheath [PALS]). T lymphocytes are most numerous closest to the arterial wall. As one moves away, more B lymphocytes are present and they cluster into follicles, as is the case in lymph nodes. Efferent lymphatics originate as closed-ended tubes in the follicles of the PALS, merge, and drain into a lymph node located in the hilus of the spleen. The PALS is referred to as the *white pulp of the spleen*. The marginal zone surrounds the PALS and covers the marginal sinus, which receives an arterial supply and contains dendritic cells, macrophages, and T and B lymphocytes. Arteries discharge into venous sinuses or a space called the *splenic cord*. These structures constitute the red pulp of the spleen. Antigen in the blood is ingested in the marginal zone and the venous sinuses. It is then presented to T lymphocytes moving through the PALS.

There are additional sites of lymphocyte aggregation in connective and parenchymal tissues of the body, particularly in the skin, the lungs, the gastrointest-inal tract, and the urogenital tract, which are in constant contact with substances external to the organism. These are collectively referred to as *MALT* or, depending on the specific organ, as *bronchial-associated lymphoid tissue* (BALT) or *gut-associated lymphoid tissue* (GALT). At all these sites, cell-mediated and humoral immune responses can be initiated. The humoral response is mediated by IgA, which helps prevent pathogen adhesion to mucosal surfaces and blocks entry.

It has been estimated that 1–2% of the lymphocyte pool recirculates each hour. Some lymphocytes appear predisposed to enter preferentially particular lymphoid compartments. Memory cells, especially, may be endowed with some ability to return to areas where antigen was first encountered. The chemokines and CAMs are essential players in the ability of leukocytes to migrate toward areas where they are needed.

XIII. IMMUNE ACTIVATION/DEACTIVATION AND MEMORY

Immune activation occurs within the context of the lymph nodes, spleen, and other lymphoid tissues. In effect, antigens that do not find their way into the secondary lymphoid tissue are not likely to activate immune responses. The interactions between APCs, T lymphocytes, and B lymphocytes that result in activation are mediated not only by the direct contact between these cells and the release of cytokines, but also by other aspects of the chemical composition of the microenvironment that surrounds the cells. The composition of the microenvironment includes even substances released by pathogens directly or from infected cells. Essentially pathogens have developed ways of evading immune responses by interfering in one way or another with the activating signals. One approach involves releasing proteins that mimic inhibitory cytokines; another is releasing soluble receptors that bind cytokines and block their activating effect.

Normal deactivation of immune activity occurs as the source of antigen or nonspecific stimulation declines in the microenvironment. Activated lymphocytes have a limited life span. Antibodies may be subject to neutralizing each other via a process whereby one antibody can bind specifically with the antigen-binding site (idiotype) of another antibody. This is referred to as *idiotype–anti-idiotype* interactions. However, such interactions are not always inhibitory and can mimic the presence of antigen. Lymphocytes also release inhibitory cytokines (e.g., IL-10), and may even direct cytotoxic activity toward each other as a way to bring about response termination. Prolonged immune activation appears to have pathological consequences.

Immune memory constitutes a form of background activation and is clearly adaptive. It permits more rapid and effective response to reinfection. How such memory is maintained, given that cells have a limited life span, may reside in continued stimulation caused by low levels of antigen remaining in the organism or some other form of stimulation (e.g., cytokines or idiotype–anti-idiotype interactions) that favors the preservation of particular clones of lymphocytes.

In this context, it should also be noted that the products of endocrine glands and the local activity of nerve terminals could modulate immune responsiveness. Much more will be said about the influence of neuroendocrine signals on the immune system in later chapters. Here, it is simply acknowledged that the interaction between cells of the immune system is clearly subject to regulation by the central nervous system. Anatomical studies have shown that both primary and secondary lymphoid tissues receive innervation from the sympathetic nervous system (SNS). The thymus appears to receive parasympathetic innervation as well. The spleen, lymph nodes, and MALT show a pattern of SNS innervation, which suggests that the process of antigen presentation leading to activation of T lymphocytes may be most affected by the activity of such nerve terminals. These findings document the structural basis for psychoneuroimmune modulation.

XIV. DEVELOPMENT OF IMMUNE FUNCTIONS

Although the basic structure of the immune system is in place at birth, its functional capability requires a number of years to develop. Pathogens are not effectively handled at birth, and some, if present early on, may be treated as self-antigen.

The structures of the immune system begin to appear shortly after conception. The lymphatic vessels begin to differentiate very early, and lymphocyte recirculation is detectable within the first trimester. MALT is central to early defense. The BALT appears mature at birth. GALT (Peyer's patches) is detectable by 11 weeks of gestation and appears mature shortly after birth, although the number of Peyer's patches increases from 60 at birth to more than 200 by early adolescence.

The inflammatory response is limited at birth because the complement components in serum are at reduced levels, 50–90% of adult values. The levels of C8 and C9 are even lower than 50%. Adult values are reached by the end of the first year of life. Neutrophil activity is significantly reduced in newborns. In particular, neutrophils exhibit diminished chemotaxis and adhesion, making them less able to deal with infection. Similarly, the macrophage system requires gradual maturation with respect to chemotaxis and cytokine production, with neonatal values ranging from 15% to 75% of adult values.

B lymphocytes are detectable at 13 weeks of gestation. At birth, IgM- and IgD-bearing B cells are most evident. The high number of IgD lymphocytes is noteworthy because they gradually disappear. B-lymphocyte proliferation is sluggish at birth, and even though all the Ig subtypes can be produced at birth, the system is not able to produce specific antibodies to some types of antigen. IgA production is not detectable until 2 weeks after birth. Secretory IgA begins to appear during the first year and reaches the adult levels in saliva and nasopharyngeal secretions at age 2–4 years of age. Serum IgA does not reach adult levels until early adolescence. IgG is the only isotype that can cross the placenta, so fetal levels are higher than in the neonate. IgG subclasses reach adult levels at various times between the ages of 5 and 12 years. The ability of T cells to recognize antigen is detectable during gestation, at the beginning of the second trimester; however, at birth, the T cells appear to exert greater suppression than at later times. The circulating T cells are predominantly naive (more than 90%), whereas in the adult the figure is approximately 60%. T-cell cytotoxicity and cytokine production are reduced in neonates. T-cell function appears adultlike by the age of four years. NK cells also have reduced cytotoxicity at birth, which does not mature until 4–5 years of age.

In various ways, immune reactivity is nonoptimal in infancy but is compensated somewhat by protection derived from maternal immune function, acquired prenatally (via the placenta) or from breast milk. The fact that the immune system is not entirely ready to operate becomes a consideration in the timing of vaccinations, which assume a competent response to antigen. A fully competent immune system may not be in place until late childhood or early adolescence, a period that coincides with the lowest mortality from a number of infectious diseases.

Changes in immune competence continue with age. Almost as soon as immune function reaches its peak, the thymus begins a process of involution. Thymic peptide levels show evidence of decline during early adulthood (20–30 years of age), and levels become undetectable in those older than 60 years. The elderly exhibit decreased lymphocyte proliferation and IL-2 production, increased IgG and IgA levels, decreased response to specific antigens, decreased delayed hypersensitivity, and decreased neutrophil function. These changes may not only reflect the aging process, but to some extent also be associated with the stress that can accompany aging in a culture that places high value on youth.

XV. MEASURES OF IMMUNE FUNCTION

Immune activity is a multifaceted process occurring at a multitude of sites throughout the body. In many ways, the immune system is as complex as the central nervous system and, thus, defies simple characterization in terms of its function. Nonetheless, a number of measures have gained popularity in the research to gain a foothold on the relation between immune function and various challenging conditions affecting the individual. This section discusses the more frequently employed measures in preparation for the later chapters dealing with psychosocial modulation of immunity.

In vivo tests involve activating the immune system by introducing a novel antigen (as in immunization) and then, after an appropriate period has elapsed, assaying for antibodies to the specific antigen in the serum of the individual. The most commonly employed assay is known as *enzyme-linked immunosorbent assay* (ELISA). Serum from the individual is incubated on a surface with the specific antigen bound to it. Antibody in serum specific for the antigen, if present, will bind to the surface. Another antibody is then introduced that binds to human antibody molecules and carries an enzyme, which catalyzes the reaction of a colorless substrate to a colored product. The intensity of the color that appears is quantitatively related to the amount of specific antibody bound to the antigen. The ELISA has a variety of other applications, which include detecting antigen (as opposed to antibody) in serum or quantifying cytokine released by activated lymphocytes. Antibody to tissue-bound antigens can be quantified using an immunofluorescence assay. This is similar to the ELISA procedure except that the starting point is actual tissue samples, exposed to serum and then incubated with antibody to human immune globulin that fluoresces in response to ultraviolet light.

Another *in vivo* assay examines the response to antigens that can be expected to have infected most individuals before testing. This method is known as the *delayed hypersensitivity skin test*. The antigens are introduced by needle puncture of the skin. They are then processed by macrophages and dendritic cells and presented to CD4$^+$ memory cells that remain from previous encounters with the antigens. The CD4$^+$ cell will release cytokines and attract neutrophils into the tissue, causing an elevated red area (inflammation) on the skin within approximately 48 hours. This is the method used to evaluate for previous infection with *Mycobacterium tuberculosis* and is known in that case as the *tuberculin test*.

A number of *in vitro* quantitative assays allow measurement of major classes of immunoglobulins (e.g., IgG, IgA, IgM, and IgE). A technique known as *nephelometry* uses antibodies against the heavy chain of major classes of antibody molecules to form immune complexes that precipitate and yield estimates of the concentration of a particular class of immunoglobulin (e.g., IgG or IgA) in the serum of an individual. Quantification of antibody specific for a given pathogen requires the ELISA technique previously described.

Quantification of leukocyte subpopulations, as well as lymphocyte subtypes, relies on techniques such as flow cytometry. The cells are prepared so they can be fed one at a time through a device with a laser that permits differential detection based on cell characteristics (e.g., size and granularity) and can sense tags (i.e., color-emitting dyes on antibodies specific for key surface molecules) such as CD3 or CD4 on the cells. Essentially, this device can separate and count cell types at a rate of 10,000 per second.

In vitro functional tests can be used to quantify the ability of lymphocytes to proliferate. The cells are incubated with tritiated thymidine, which is incorporated into the DNA of dividing cells, and are stimulated to divide using mitogens (i.e., nonspecific substances that induce mitosis, such as phytohemagglutinin, concanavalin A, and pokeweed mitogen). The amount of radioactivity detected provides an indication of the amount of cell division (proliferation) in the culture. It is also possible to assess NK-cell activity by incubating the NK cells with target cells that have been labeled intracellularly with radioactive chromium. The ability of NK cells to lyse the target cells causes a release of radioactive chromium and provides a measure of NK-cell cytotoxicity.

These approaches to quantifying immune function or responsiveness are limited because in humans, the variation that has been observed is well within the normal range and possibly not clinically significant. Moreover, except for skin testing, the measurements are based on peripheral blood samples, which may not adequately reflect what is transpiring within the lymphoid tissues or at sites of infection or malignant transformation.

XVI. CONCLUDING COMMENTS

As noted earlier in this chapter, the immune system is complex. However, the notion of pathogen penetration can be used to present an integrated view of immune defenses (Figure 3.15). Pathogens pose a danger only if they enter the body; thus, the first line of defense is a variety of physical and molecular barriers to entry on the surfaces of the body that are exposed to the external world. Once a pathogen gets through such barriers, it can remain in the intercellular spaces or enter the host's cells. Extracellular pathogens or their products (e.g., toxins) can be detected by nonspecific defenses (e.g., phagocytes, complement, and NK lymphocytes). Essentially, such pathogens have membrane characteristics or structures that the leukocytes can recognize as foreign. However, as noted earlier in this chapter, disposal of such pathogens can be enhanced by the activation of specific immune mechanisms (i.e., presentation of antigen to helper T cells), which in turn activates specific antibody production. The binding of specific antibody to pathogens or their toxins enhances the ability of phagocytes, complement, and even NK cells to dispose of the invader. Moreover, if antibody of the IgA variety is produced, then pathogen entry through mucosal surfaces will be even more effectively blocked.

Intracellular pathogen (e.g., virus) or abnormal transformation of cells (i.e., malignancy) poses additional challenges. This is when specific immunity, especially cell-mediated effector activity, comes to the forefront. The HLA molecules of the cells must be inspected for the presence of self-antigen or foreign antigen. NK cells will destroy cells if the HLA class I molecules are not evident. CD8$^+$ T cells will destroy cells if HLA class I molecules are expressed and display a foreign antigen. This ensures that the cytotoxic substances released are highly targeted and spare adjacent cells. Activation of helper T cells is essential to an overall targeting and enhancing of immune responsiveness. In effect, if an antigen produces T-cell activation, the effort to get rid of it will be multifaceted and enhanced. Termination of responses may be achieved via a variety of internal (immune system proper) mechanisms or may involve shifts in signals arising from other tissues (endocrine and neural) that are capable of modulating immune function. Finally, given the complex and probabilistic nature of how immune responses are mounted (i.e., antigen and specific receptors must meet before it is too late), significant variation must be expected over time, as a function of situation, and across individuals, but at any point in time, only a small fraction of the possible unique TCRs are actually expressed and available for antigen recognition. These issues are underscored in subsequent chapters in this book.

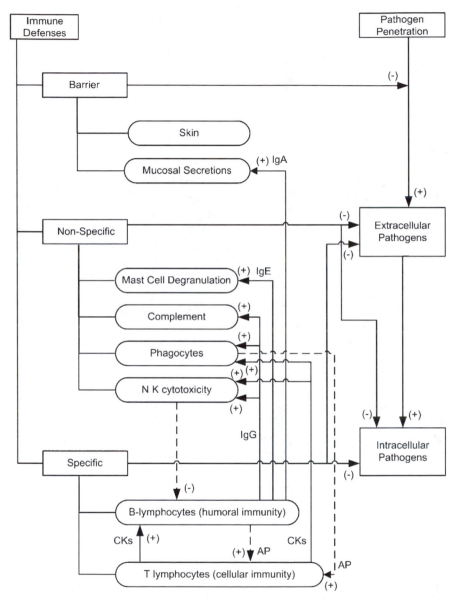

Figure 3.15 General organization of immune defenses. AP, antigen presentation; CKs, cytokines; IgG, immunoglobulin G; NK, natural killer.

XVII. SOURCES

Cohen, I.R. (2000), *Tending Adam's Garden: Evolving the Cognitive Immune Self*, San Diego, Academic Press.

Cruse, J.M. & Lewis, R.E. (1999), *Atlas of Immunology*, Boca Raton, Fla, CRC Press.

English, B.K., Schroeder, H.W., Jr, & Wilson, C.B. (2001), "Immaturity of the fetal and neonatal immune system," vol 1, in Rich, R.R., Fleisher, T.A., Shearer, W.T., et al. (eds) *Clinical Immunology: Principles and Practice*, 2nd ed, London, Mosby.

Fagarasan, S. & Honjo, T. (2000), "T-Independent Immune Response: New Aspects of B Cell Biology," *Science*, 290, pp. 89–92.

Hayday, A. & Viney, J.L. (2000), "The Ins and Outs of Body Surface Immunology," *Science*, 290, pp. 97–100.

Lanzavecchia, A. & Sallusto, F. (2000), "Dynamics of T Lymphocyte Responses: Intermediates, Effectors, and Memory Cells," *Science*, 290, pp. 92–97.

Matzinger, P. (2002), "The Danger Model: A Renewed Sense of Self," *Science*, 296, pp. 301–305.

Medzhitov, R. & Janeway, C.A., Jr (2002), "Decoding the Patterns of Self and Nonself by the Innate Immune System," *Science*, 296, pp. 298–300.

Pabst, H.F. & Kreth, H.W. (1980), "Ontogeny of the Immune Response as a Basis of Childhood Disease," *Journal of Pediatrics*, 97, pp. 519–534.

Paul, W.E. (ed) (1999), *Fundamental Immunology*, 4th ed, Philadelphia, Lippincott–Raven Press.

Rabin, B.S. (1999), *Stress, Immune Function, and Health: The Connection*, New York, Wiley-Liss.

Ravetch, J.V. & Lanier, L.L. (2000), "Immune Inhibitory Receptors," *Science*, 290, pp. 84–89.

Rich, R.R., Flisher, T.A., Shearer, W.T., et al. (eds) (2001), *Clinical Immunology Principles and Practice*, 2nd ed, vol 1, London, Mosby.

Roitt, I.M. & Delves, P.J. (2001), *Roitt's Essential Immunology*, 10th ed, Oxford, Mass, Blackwell Science.

CHAPTER 4

Endocrine-Immune Modulation

I. Introduction 58
II. Endocrine System 59
   A. Hypothalamus 59
   B. Pituitary Gland 60
   C. Other Glands 62
III. Cytokines, Hormones,
and Their Receptors 63
IV. Anterior Pituitary Hormones
and Immune Function 64
   A. Growth Hormone 64
     1. Regulation of Release 64
     2. Modulation of Immune Function 65
   B. Prolactin 66
     1. Regulation of Release 66
     2. Modulation of Immune Function 67
   C. The Pituitary–Adrenal Axis: POMC Peptides
     (ACTH, β-Endorphin)
     and Glucocorticoids 67
     1. Regulation of Release 67
     2. Modulation of Immune Function 68
   D. The Pituitary–Gonadal Axis: Gonadotropins
     and Gonadal Steroids 70
     1. Regulation of Release 70
     2. Modulation of Immune Function 71
   E. The Pituitary–Thyroid Axis: Thyrotropin and
     Thyroid Hormones 72
     1. Regulation of Release 72
     2. Modulation of Immune Function 72
V. Posterior Pituitary Hormones
and Immune Function 73
   A. Arginine Vasopressin 73
     1. Regulation of Release 73
     2. Modulation of Immune Function 73

     B. Oxytocin   73
        1. Regulation of Release   73
        2. Modulation of Immune Function   74
 VI. **Other Hormones and**
      Immune Function   74
     A. Insulin   74
        1. Regulation of Release   74
        2. Modulation of Immune Function   74
     B. Parathyroid Hormone   74
        1. Regulation of Release   74
        2. Modulation of Immune Function   75
     C. Melatonin   75
        1. Regulation of Release   75
        2. Modulation of Immune Function   75
VII. **Thymus Gland   76**
     A. Pituitary Regulation of Thymus   76
     B. Thymus Regulation of Pituitary   76
     C. Hormones and Thymocyte
        Development   77
VIII. **Concluding Comments   77**
 IX. **Sources   79**

I. INTRODUCTION

The preceding chapter provided an overview of immune defenses and underscored that immune responses require intercommunication by a variety of cell types. It was alluded that molecular signals capable of enhancing or suppressing immune responsiveness not only originate within the immune system but also can arise from other systems. In effect, evidence suggests that multiple signals must converge for triggering, sustaining, and terminating an effective immune response.

This chapter focuses on the role of the endocrine system in the modulation of immune function. The presentation begins with an overview of the major structures that compose the endocrine system and a catalog of the major hormones released by the endocrine glands. It will be apparent that the release of hormones by the endocrine glands is to a significant extent under neural control, is subject to negative feedback, and is influenced by other hormones. In addition, substances that were first discovered in immune cells (e.g., cytokines) and that can alter endocrine activity are now known to be released also by pituitary cells, endothelial cells, glial cells, and even neurons. Nonetheless, at least some of the cytokine effects on hormone release must arise at distal sites because they appear to be mediated by the peripheral nervous system. Moreover, leukocytes have been shown to produce pituitary hormones and various other peptides and could in principle influence neuroendocrine activity. As these potential reciprocal influences between the endocrine and immune systems are described, the abundance of pathways for neuroimmune communication will become obvious. At the same time, the sheer complexity of the emerging circuitry precludes the notion that a simple model of ultimate hormonal immune effects can actually be found.

II. ENDOCRINE SYSTEM

A general overview of the endocrine system, its hormones, and some of their major effects will serve as the backdrop for the discussion of how hormones modulate the function of the immune system in response to activity arising within the hypothalamus.

A. HYPOTHALAMUS

The hypothalamus lies at the base of the brain. Neurons located within the various hypothalamic nuclei are directly involved in the release of the pituitary hormones, which, in turn, affect the activity of the other endocrine glands. Thus, one facet of the influence of the central nervous system over immune function occurs via the endocrine system.

Neurons that are hypophysiotropic (i.e., stimulate the pituitary) are widely distributed within the hypothalamus, but all project to the median eminence, where they secrete their specific "releasing hormones" into the portal capillaries. The various releasing hormones then flow to the anterior pituitary cells. These cells are, in turn, the source of the "stimulating hormones," which ultimately regulate the activity of several of the major endocrine glands. Two other groups of neurons directly project to the posterior pituitary and are the source of the hormones oxytocin and arginine vasopressin (AVP), released by the posterior pituitary.

Neurons containing the various releasing hormones are not confined to the hypothalamus, but they are also widely distributed throughout the nervous system. Moreover, nonneural cells have also been found to contain the releasing hormones. The significance of this observation, which has become quite common, is that the same molecule can perform different functions within different tissues. In other words, corticotropin-releasing hormone (CRH) is not restricted to inducing the release of corticotropin from cells; it can also function as a modulator of neural activity in other brain regions or even as a "cytokine" in lymphoid tissue.

Releasing hormones secreted into the portal capillaries of the anterior pituitary regulate the release of adrenocorticotropic hormone (ACTH, corticotropin), follicle-stimulating hormone (FSH), luteinizing hormone (LH), growth hormone (GH), thyroid-stimulating hormone (TSH), and prolactin (PRL). The neurons that are the source of the releasing hormones are distributed among several hypothalamic nuclei. However, the paraventricular nucleus (PVN) appears to be particularly prominent, especially with respect to immune modulation. It contains cell bodies for neurons that secrete the posterior pituitary hormones (AVP, oxytocin), as well as neurons that secrete CRH, thyrotropin-releasing hormone (TRH), GH-releasing hormone (GH-RH,) and somatotropin release–inhibiting hormone (i.e., somatostatin, a hormone that inhibits the release of GH). In addition to its role in endocrine function, the PVN contains neurons that project to brainstem nuclei and to spinal cord neurons that are part of the autonomic nervous system.

It is important to emphasize that there is a high degree of interconnectivity among hypothalamic nuclei that have been implicated in endocrine regulation. These nuclei also receive substantial input from other brain regions, particularly from the brainstem and limbic structures. A wide range of neurotransmitters and neuropeptides influence the activity of these cells. Even macromolecular substances that are unable to penetrate the blood–brain barrier (BBB) appear capable of influencing hypothalamic activity via circumventricular structures, which lack a BBB, such as the organum vasculosum of the lamina terminalis (OVLT), the subfornical organ (SFO), and the area postrema (AP). Computation within the hypothalamus can simultaneously result in output directed toward the limbic system (e.g., amygdala, hippocampus, and septal area), the sensory systems (e.g., thalamus), the autonomic nervous system (e.g., brainstem and spinal cord), and the endocrine system (e.g., pituitary). Modulation of the immune system emerges as one of the consequences of such computation.

B. Pituitary Gland

The pituitary gland, or hypophysis, lies at the base of the brain, just behind the eyes, and it is directly connected to the hypothalamus by the pituitary stalk. It is composed of an anterior lobe, an intermediate zone (vestigial in humans), and a posterior lobe.

The anterior pituitary is the source of key hormones that serve to stimulate many of the other endocrine glands (Figure 4.1). The principal hormones secreted by the anterior pituitary include the following: (1) GH, a protein that serves to stimulate bone growth and regulate aspects of metabolism, (2) PRL, a protein whose structure resembles that of GH and serves to stimulate production and secretion of breast milk and to produce uterine contractions, (3) FSH, a glycoprotein that stimulates production of ova and sperm by the gonads, (4) LH, a glycoprotein that stimulates the ovaries and testes to release gonadal steroids, (5) TSH, a glycoprotein that stimulates the release of thyroid gland hormones, and (6) the cleavage

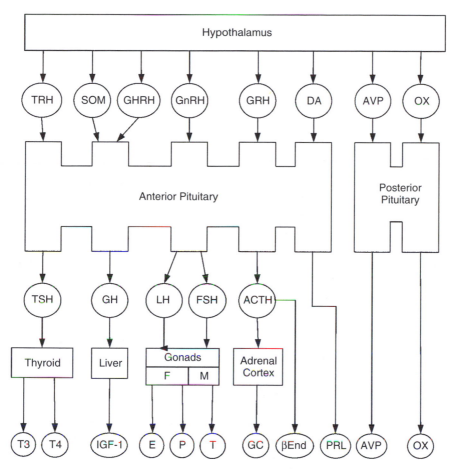

Figure 4.1 Hypothalamic–pituitary–endocrine gland axes. ACTH, adrenocorticotropic hormone; AVP, arginine vasopressin; β-end, β-endorphin; CRH, corticotropin-releasing hormone; DA, dopamine; FSH, follicle-stimulating hormone; GC, glucocorticoids. GH, growth hormone; GH-RH, growth hormone–releasing hormone; GnRH, gonadotropin releasing hormone; IGF, insulin-like growth factor; LH, luteinizing hormone; OX, oxytocin; P, progesterone; PRL, prolactin; SS, somatostatin; T, testosterone; T_3, triiodothyronine; T_4, thryoxine; TRH, thyrotropin-releasing hormone; TSH, thyroid-stimulating hormone.

products of the proopiomelanocortin (POMC) protein, which include ACTH, a peptide that stimulates glucocorticoid release by the adrenal cortex. ACTH is co-released with β-endorphin, a peptide that has analgesic properties. In humans, α-melanocyte–stimulating hormone (α-MSH), another POMC cleavage product, is not detectable in the circulation, except during pregnancy.

The posterior pituitary acts as a conduit for the secretion of peptides that arise from neurons located within the hypothalamus. The principal peptides released by the posterior pituitary include oxytocin, which stimulates contraction of the uterus and the mammary glands, and AVP (a.k.a., antidiuretic hormone) a peptide that promotes retention of water by the kidneys.

C. OTHER GLANDS

The thyroid gland is located in the throat area. It is stimulated by TSH to release triiodothyronine (T_3) and thyroxine (T_4), amines that enhance metabolic activity. The thyroid also releases calcitonin, a peptide that acts to lower the blood calcium level. The parathyroid glands are embedded in the thyroid gland and release parathyroid hormone (PTH), a peptide that raises the blood calcium level (i.e., counteracts calcitonin).

The islets of Langerhans are small clusters of cells accounting for 1–2% of pancreatic tissue, which are considered a part of the endocrine system. These cells are the source of insulin and glucagon, proteins that serve to regulate (i.e., decrease and increase, respectively) the blood glucose level.

The adrenal glands are located above the kidneys. They are composed of a cortex (outer) and a medulla (inner) region. The adrenal medulla is more appropriately regarded as part of the sympathetic nervous system, although it has endocrine properties as well. It releases epinephrine and norepinephrine, as well as several peptides in response to neural signals arriving via preganglionic sympathetic nerve fibers. The release of epinephrine and norepinephrine into the circulation raises blood glucose level, increases metabolic activity, and constricts certain blood vessels to redirect blood flow. The effects of epinephrine and norepinephrine (i.e., catecholamines) on immune function are presented in Chapter 5, because in this chapter the concern is with the glucocorticoid hormones released by the adrenal cortex in response to ACTH. The glucocorticoids, along with the catecholamines, are central to the stress response (i.e., mobilization of bodily functions and resources to threat [or more generally the unknown] to sustain fight or flight responses). The adrenal cortex also releases mineralocorticoids, which promote the reabsorption of sodium and excretion of potassium by the kidneys.

The gonads (ovaries and testes) respond to FSH and LH. In the male, the testes release testosterone, which promotes the development and maintenance of male secondary sex characteristics. In the female, the ovaries release estradiol, which promotes the development and maintenance of female secondary sex characteristics. Progesterone, also released by the ovaries, promotes growth of the lining of the uterus.

The pineal gland, located above the midbrain region and nestled between the cerebral hemispheres, is the source of melatonin, an amine that plays a key role in the regulation of circadian rhythms in response to the light–dark cycle.

The thymus gland, one of the primary lymphoid tissues, is also considered part of the endocrine system. The thymus is located just above the heart and is the source of a family of peptides (e.g., thymopoietin, thymulin, and thymosin) that not only promote T-cell maturation locally but also appear to have distal effects on the pituitary and on neural elements that are part of the circuitry regulating the endocrine glands. This dual role serves to illustrate how our approach to categorizing tissues into systems cannot avoid overlap because in essence, the organism is an integrated system.

III. CYTOKINES, HORMONES, AND THEIR RECEPTORS

Although the focus of this chapter is on modulation of immune function by the endocrine system, in reality these systems modulate each other (i.e., there is bidirectional communication).

The least complicated version of such intercommunication is characterized by lymphoid tissue expressing specific receptors for hormones that are exclusively produced within the endocrine glands. The feedback channels would rely on substances released exclusively by lymphoid tissues and leukocytes (e.g., cytokines) for which specific receptors would be expressed only in particular neural pathways and/or hormone-releasing cells. Cytokine receptors have been found in the brain and the pituitary, whereas hormone receptors are expressed by leukocytes and lymphoid tissues. However, as noted earlier, the manner in which these systems intercommunicate is greatly complicated by evidence that anterior pituitary cells, as well as neurons and glial cells, produce and release cytokines. Among the cytokines that have been found are IL-1, IL-1RA, IL-2, and IL-6. They influence neural activity and the growth and function of the hormone-releasing cells. Thus, whether the receptors are there to mediate local effects (autocrine/paracrine) or to receive signals arising at distal sites (endocrine) remains unsettled. A similar complication arises from the discovery that most if not all of the pituitary hormones appear to be synthesized and released by leukocytes and other lymphoid tissues. For instance, it has been shown that at least *in vitro* T lymphocytes can be stimulated to produce ACTH, endorphins, TSH, GH, PRL, and other substances such as insulin-like growth factor-1 (IGF-1), a protein that appears to mediate GH effects on various tissues including the immune system. B lymphocytes, macrophages, and the thymic epithelium are similarly able to produce hormones.

Given the aforementioned observations, modulation of lymphocyte activity by a particular hormone (e.g., GH) could, therefore, arise from the lymphocyte itself (autocrine effect), from a nearby lymphocyte or macrophage (paracrine effect), or from the anterior pituitary (endocrine effect). In essence, from the point of view of receptors on a given lymphocyte, the source of specific molecules acting on it is indeterminate unless one assumes that there is some sort of "tagging" due to differences in ligand molecular structure that are dependent on tissue of origin.

IV. ANTERIOR PITUITARY HORMONES AND IMMUNE FUNCTION

Studies of the effect of pituitary removal (hypophysectomy) have been conducted for some time and predate the discovery of leukocyte-derived "pituitary" hormones. In retrospect, these studies help to put in perspective the functional importance of pituitary-derived hormones.

Studies conducted during the early 1900s demonstrated that removal of the pituitary gland had profound effects on the organism. Animals without pituitaries have difficulty surviving. With careful control of the postoperative environment, the survival of these animals can be prolonged. The animals are then found to have problems regulating body temperature, to exhibit learning deficits, and particularly interesting in this context, they appear more susceptible to infection. A general immunodeficiency is observed in the hypophysectomized animals. The pituitary hormones, directly and through their effects on other glands, play a crucial role in hematopoiesis (generation of blood cells by bone marrow) and in maintaining the function of the thymus. The spleen and lymph nodes atrophy in hypophysectomized animals. Suppressive effects have also been found on antibody production, natural killer (NK) cell activity, and lymphocyte proliferation.

Administration of GH or PRL is capable of restoring some of the immune functions of hypophysectomized animals. This is not the case if the animals are treated with ACTH or TSH. In fact, ACTH antagonizes the restoration produced by GH or PRL. These findings appear consistent with the interpretation that GH and PRL serve as essential factors for immune function. It is also clear from these studies that even though leukocytes can release both GH and PRL, such a source is not sufficient to sustain normal immune function. Leukocyte-derived hormones may operate in a localized manner, perhaps to sustain activity briefly, but cannot compensate for the loss of the pituitary.

A. GROWTH HORMONE

1. Regulation of Release

The release of GH (somatotropin) from the anterior pituitary cells, known as *somatotrophs*, is influenced by a wide range of factors. As the name implies, GH-RH, secreted by cells located in the arcuate nucleus of the hypothalamus, promotes GH release. In contrast, somatostatin, which is secreted by cells located in the preoptic area and the anterior hypothalamus, inhibits the release of GH. So far, this is the only case of a well-characterized antagonistic hormone-release duo, although there is evidence that eventually other such hormone duos will be discovered. These hormones act directly on the somatotrophs to exert their effects on release. Moreover, they also antagonize each other more directly in that GH-RH stimulates somatostatin neurons, whereas somatostatin neurons inhibit themselves and GH-RH neurons. In addition, these neurons receive input from various other brain structures and their activity is affected by the release of multiple neurotransmitters (e.g., γ-aminobutyric acid [GABA] and norepinephrine) and neuropeptides (e.g., vasoactive intestinal peptide [VIP], galanin, and glucagon). The somatostatin neurons also appear responsive to various metabolic inputs; in particular, hypoglycemia and increased insulin serve to inhibit somatostatin neuron activity. Free fatty acid levels and glucocorticoids stimulate somatostatin release. At the same time, glucocorticoids enhance the release of GH by anterior pituitary cells. In effect, glucocorticoids not only promote the inhibition of GH release via somatostatin, but also create conditions that enhance GH release when the inhibition is overridden.

GH release is stimulated by thyroid hormones and testosterone. It is inhibited by estradiol. GH regulates its own release both at the hypothalamic level and at the level of the pituitary. It inhibits somatotrophs and GH-RH neurons while stimulating somatostatin neurons. Finally, GH stimulates the production of IGF-1 primarily from the liver. IGF-1 mediates the growth effects of GH and provides negative feedback in parallel to GH to the same areas of the hypothalamus and the pituitary, adding yet another layer of regulatory control.

Thus, the convergence of multiple influences, acting via a variety of intracellular messengers, ultimately serves to shift the somatotrophs into a more or less inhibited state. It is as if other bodily systems cast a vote for or against the release of GH on a case-by-case basis. It is probably true that some inputs (votes) carry more weight than others. The fact that GH release is predominantly pulsatile and nocturnal suggests that favorable circumstances are short lived and related to the arousal state of the organism. The picture that has been sketched here is likely, at least partially, to fit what occurs in the case of other hormones.

2. Modulation of Immune Function

GH has powerful effects on immune function. One of the earliest observations concerning the effect of GH on the immune system was its ability to increase the size of lymphoid tissues, particularly the thymus. As noted earlier in this chapter, the adverse effects of hypophysectomy on immune responsiveness can be ameliorated by GH. GH receptors have been found for all hematopoietic lineages (B cells, macrophages, T cells, and granulocytes). High-affinity receptors for GH are most evident on B cells and macrophages and least evident on T cells. However, the action of GH on lymphoid tissue (as on other tissues) appears to be mediated in great part by its ability to promote the release of IGF-1, which functions as a potent growth factor for hematopoiesis. IGF-1 appears more proximal in the effect chain, because blocking the action of IGF-1 prevents the effects of GH.

GH augments lymphocyte proliferation and cytotoxicity by CD8$^+$ T cells and NK cells. GH can enhance the maturation of granulocytes in the presence of normal signals for their maturation (e.g., IL-3 and granulocyte–macrophage colony-stimulating factors [GM-CSFs]). GH may have important effects on the ability of macrophages to digest pathogens. GH increases the macrophages' production of superoxide anion (O_2^-), which acts to lyse ingested pathogens. Again, these effects depend on the production of IGF-1, most likely from macrophages.

Antigen has been found to increase GH release. Malignancy (a source of antigen) has also been found to increase GH concentration in plasma. GH secretion peaks around puberty. The subsequent decline in GH secretion is associated with a decrease in the plasma concentration of IGF-1 and with thymic involution. Thymic involution can be reversed by GH, which also modulates thymulin production.

Overall, GH appears to have immune function–enhancing properties. However, the immune-enhancing effects of exogenous GH are most evident in hypophysectomized animals. Moreover, it is noteworthy that GH-deficient children are not more susceptible to infection, although some indices of immune function appear depressed in such children. These observations suggest that other hormones (e.g., prolactin) may compensate for the absence of GH and that low levels of GH may be adequate to sustain effective immune responses.

B. Prolactin

1. Regulation of Release

The release of PRL from the anterior pituitary mammotrophs is subject to various influences. Dopamine-containing neurons located in the tuberoinfundibular region of the hypothalamus exert a major inhibitory effect over PRL release. This effect is mediated via D_2 receptors. PRL increases the activity of the dopaminergic neurons and downregulates its own release. TRH and VIP are potent stimuli for PRL release. Endogenous opioids, by inhibiting dopaminergic neurons, increase PRL release. Estrogens increase PRL release, whereas glucocorticoids inhibit PRL release. Proinflammatory cytokines (e.g., IL-1) inhibit PRL release, whereas thymic peptides increase its release.

As noted earlier in this chapter, the level at which these modulatory effects occur can vary between the hypothalamus and the pituitary. In other words, the primary effects can be on the dopaminergic neurons, the pituitary cells, or both. Moreover, the source of the modulating molecule can be the cell itself, a nearby cell, or a more distant cell. For instance, the IL-1 effect on PRL release may originate from hypothalamic neurons, neuroglia, pituitary cells, more distally located macrophages, or possibly all of these.

2. Modulation of Immune Function

PRL was originally named in the 1930s because it could stimulate lactation. However, early research had also demonstrated that PRL deficiency was associated with impaired immune responses. PRL shares structural similarities with GH, which facilitates immune responsiveness, although PRL has its own distinct receptor. Thus, PRL has been regarded as immunopermissive.

Drugs that stimulate D_2 receptors and inhibit PRL release have been found to suppress antibody formation in response to specific antigens and to suppress the delayed hypersensitivity response. These effects can be counteracted by administering exogenous PRL or GH. Dopamine agonists have also been found to interfere with T-cell proliferation and block the activation of macrophages by T cells. PRL appears to be critical in facilitating the early stages of macrophage activation. Drug-induced inhibition of PRL release increases the lethality of infections.

Dopamine antagonists such as the neuroleptics increase the release of PRL and enhance lymphocyte proliferation. They are also able to counteract the immunosuppressive effects of glucocorticoids (see later discussion) and synthetic drugs such as cyclosporine. In fact, cyclosporine, a potent immunosuppressive agent frequently used in clinical practice to prevent rejection of transplanted organs, may be a PRL receptor antagonist. PRL is necessary for the expression of the IL-2 receptor and enhances the production of interferon-γ (IFN-γ) by mononuclear leukocytes. PRL may potentiate Th1 activity and, thus, cell mediated immune responses. It is noteworthy that increased PRL levels have been consistently found in some animal models of autoimmune disorders and in patients afflicted with such disorders. Increased PRL has also been observed to be an early indication of impending rejection of transplanted tissue.

C. THE PITUITARY–ADRENAL AXIS: POMC PEPTIDES (ACTH, β-ENDORPHIN) AND GLUCOCORTICOIDS

1. Regulation of Release

The major signal regulating the pituitary–adrenal axis is provided by the release of CRH from neurons located within the PVN of the hypothalamus. CRH has been detected in other brain regions and other organs, where it serves a variety of functions, but when it is released into the portal blood vessels, it causes the release of ACTH from corticotrophs in the anterior pituitary. It also stimulates the synthesis of POMC, the precursor to ACTH and other important peptides. Evidence indicates that ACTH and other cleavage products of POMC (e.g., β-endorphin) are co-released in response to CRH.

The release from corticotrophs of POMC-derived peptides is complex because CRH release is subject to modulation by a wide range of neurotransmitters, neuromodulators, hormones, peptides, and cytokines. AVP augments the CRH effect on the pituitary, whereas opiates appear to blunt the CRH effect. Neurotransmitters such as acetylcholine, norepinephrine, and serotonin increase CRH release, whereas GABA and the gaseous neuromodulator nitric oxide decrease CRH release. Substance P decreases CRH release, whereas neuropeptide Y enhances it. Estrogen enhances CRH release and diminishes the glucocorticoid negative-feedback effect. CRH release has also been observed to increase in response to various cytokines (e.g., IL-1α, IL-1β, IL-6, and TNF-α). Moreover, as noted earlier, AVP augments the CRH effect on the corticotrophs. As a result, the release of CRH is further subject to modulation by molecules involved in blood pressure regulation, such as angiotensin II, which increases release of AVP and atrial natriuretic peptide, which decreases AVP release.

2. Modulation of Immune Function

ACTH decreases antibody production in response to antigen, interferes with macrophage-mediated tumoricidal activity, and suppresses cytokine (e.g., IFN-γ) production. In addition, ACTH increases and decreases lymphocyte proliferation, depending on dosage and other factors operating prior to stimulation with mitogens. As noted earlier in this chapter, CRH has also been detected in other tissues such as the placenta, the gastrointestinal tract, the thymus, the reproductive tract, and the skin. Peripheral CRH is capable of inducing POMC expression in leukocytes in a manner similar to that which has been observed in the pituitary. Furthermore, CRH can promote the release of cytokines (IL-1, IL-6) from monocytes, which, in turn, promote the production of POMC-derived peptides by acting on lymphocytes. These actions can be blocked by glucocorticoids. Thus, it is important to keep in mind that leukocytes can mimic hypothalamic–pituitary–adrenal (HPA) influences locally; in other words, they can produce CRH, induce POMC, secrete its products, and affect immune activity within circumscribed regions of the organism. POMC-derived peptide levels in the circulation do not appear to be high enough to cause distal immune effects. Direct effects by ACTH, β-endorphin, and α-MSH on immune function appear mostly due to their production in the periphery.

β-Endorphin and other opioid peptides appear capable of suppressing or enhancing immune responsiveness. For instance, antibody production in response to antigen stimulation can be modulated in a baseline-dependent manner. β-Endorphin increased antibody production when baseline response was low and decreased it when it was high. Thus, opioid peptides have been regarded as fine-tuning immunomodulators. They also appear to have inhibitory effects on chemotaxis. α-MSH is generally an anti-inflammatory antipyretic peptide that may operate primarily by antagonizing the actions of cytokines such as IL-1.

Glucocorticoids are probably the most extensively studied hormones with respect to their immunomodulating effect. As early as 1855, Addison observed that insufficient adrenal function was associated with an excess of white blood cells. Adrenalectomized animals were noted to have an enlarged thymus. Selye demonstrated in the mid-1930s that stress (noxious stimulation) caused adrenal enlargement and atrophied the thymus. Over the next 2 decades, adrenal hormones were first thought to enhance immune responses (i.e., they increased antibody titers) and later were considered to downregulate immune responses (i.e., decrease inflammation). By the 1970s, glucocorticoids were more generally viewed as serving to shut off activated immune responses to prevent the damage of an unregulated immune response.

Glucocorticoid effects are not simply unidirectional (i.e., suppressive). Suppressive effects appear to require prolonged exposure to relatively high levels; otherwise, either no effect or enhanced responsiveness has been observed. The level of corticosteroid-binding globulin in the circulation influences the effect of glucocorticoids by making them less available to intracellular receptors. Suppressive effects occur when glucocorticoid binding to intracellular receptors (of which there are on average 10^3–10^4 per cell) occurs in sufficient numbers to alter the expressions of specific genes. Actually, the number of receptors depends on cell type, stage of the cell cycle, state of activation, and glucocorticoid exposure. More receptors are expressed when the level of glucocorticoid decreases. Glucocorticoids then have a general inhibitory effect on many mediators of immune activation, particularly proinflammatory cytokines, such as IFN-γ, GM-CSF, IL-1, IL-2, IL-3, IL-6, IL-12, and TNF-α, as well as other mediators of inflammation (e.g., histamine, prostaglandins, and leukotrienes). Glucocorticoids do not appear to affect IL-4 and, thus, can promote a shift from Th1 (cell-mediated immunity) to Th2 (humoral-mediated immunity) predominance. Glucocorticoids prevent the accumulation of leukocytes at sites of inflammation and more generally affect leukocyte traffic. This action may be mediated in part by lipocortin-1, a protein that is produced by leukocytes in response to glucocorticoid and then binds to receptors on leukocytes that downregulate entry into sites of inflammation. Thus, it appears that through actions on the expression of cytokines, as well as adhesion molecules and receptors on leukocytes, the glucocorticoids alter the response of leukocytes in a way that diminishes antigen presentation and inflammation but promotes antibody synthesis.

D. THE PITUITARY–GONADAL AXIS: GONADOTROPINS AND GONADAL STEROIDS

1. Regulation of Release

In primates, including humans, neurons that release gonadotropin-releasing hormone (Gn-RH, a.k.a., *luteinizing hormone–releasing hormone*) are located in the medial basal hypothalamus and the arcuate nucleus. Gn-RH stimulates the release of LH and FSH. Release of Gn-RH is, in turn, regulated by feedback from the gonadal steroids and further modulated by a variety of neurotransmitters and neuromodulators impinging upon the Gn-RH neurons.

Gn-RH can inhibit its own release and regulate the expression of its receptor. In humans, dopamine appears to inhibit Gn-RH release. GABA has an inhibitory effect as well. Excitatory amino acids (glutamate, N-methyl-D-aspartate) appear to facilitate Gn-RH release. Regulation of Gn-RH by astroglial cells has been hypothesized and thought to be mediated for the release of prostaglandins and nitric oxide. Gonadal hormones, estradiol and progesterone in females and testosterone in males, inhibit the release of Gn-RH. Prolactin can inhibit the expression of Gn-RH receptors, and opiates can also modify the expression of Gn-RH receptors.

As was true for CRH, Gn-RH is found in neural pathways other than those directly involved in the release of gonadotropins. Neurons that release Gn-RH have been found to project to the hippocampus and the midbrain. Gn-RH has also been found to be widely distributed and not simply localized in neurons. It can be found in the ovary, testes, placenta, mammary glands, prostate, pancreatic islets, and the thymus. It appears to serve local regulatory functions.

Receptors for Gn-RH are found in leukocytes and other lymphoid tissue. The development and function of the hypothalamic–pituitary–gonadal axis is profoundly influenced by the presence of an intact immune system (particularly the thymus), and conversely, normal gonadal function appears necessary to sustain well-regulated immune function. Thymic peptides (thymosin fraction 5 and one of its constituents, thymosin β_4) can directly modulate gonadotropin release by acting on Gn-RH neurons within the hypothalamus. Cytokines (e.g., IL-1) are able to inhibit Gn-RH release and, thus, decrease LH output. This effect may be ultimately mediated by nitric oxide. Lastly, Gn-RH cells may also release cytokines such as IL-6, which appear, in turn, to stimulate Gn-RH release. The interdependence of gonadal and immune function is particularly noteworthy because these systems are crucial to the continuity of life.

The dynamics of Gn-RH regulation differs according to gender. Gonadal steroid feedback is negative in males, at the level of both the hypothalamus and the pituitary. In contrast, females exhibit both positive and negative feedback. During pregnancy, Gn-RH is released within the placenta, particularly early in gestation, which stimulates the placenta to secrete human chorionic gonadotropin (hCG) hormone. Human chorionic gonadotropin release, in turn, appears to be further regulated by other hormones, peptides, and cytokines in a manner analogous to what occurs at the hypothalamic–pituitary level. The resulting elevation of gonadal steroid hormones appears necessary to sustain pregnancy, largely by modifying immune responsiveness.

2. Modulation of Immune Function

The immune effects of gonadotropins (FSH, LH) appear largely mediated by their effects on the gonadal steroids. Evidence of gender differences in immune responsiveness is clear. In general, females are immunologically more responsive than males. Perhaps as a consequence, they have a higher susceptibility to auto-immune disorders and allergies. It is also true that during pregnancy immune responsiveness must be curtailed to prevent immunological attack on the fetus. The latter appears to involve a downregulation of cell-mediated immunity. In other words, the Th1–Th2 balance is shifted in favor of Th2, production of IL-2 and IFN-γ is decreased, whereas IL-4 and IL-10 are increased.

The presence of gonadal steroid receptors on leukocytes has been difficult to demonstrate. There is some evidence for the existence of cytoplasmic estrogen receptors in CD8$^+$ lymphocytes. Gonadal steroids, especially testosterone, may act primarily through effects on the thymic epithelium, where androgen receptors have been found. Gonadal steroids may exert additional effects indirectly by modulating the activity of the HPA axis.

Testosterone has consistently appeared to be immunosuppressive and can increase susceptibility to infection. Estrogens appear to depress cell-mediated immunity and NK cell activity and can increase antibody responses to T-cell–dependent antigens. Effects to both augment and reduce phagocytosis by macrophages have been noted in response to progesterone. The relationship between estrogens and immune responsiveness appears complicated. On the one hand, females are more immunologically responsive, and on the other, blockade of estrogen enhances immune responsiveness. This type of relationship has prompted the suggestion that estrogen may serve to restrain a more naturally reactive immune system in females. It is of interest in this regard that estrogen appears to enhance the activity of the HPA axis through effects on CRH release and blockade of the glucocorticoid-inhibitory feedback. In contrast, testosterone does not potentiate the HPA axis. It blunts the response to stress and enhances glucocorticoid-inhibitory feedback. Moreover, as noted earlier in this chapter, testosterone may have much more direct immunosuppressive effects that could protect against autoimmune disorder. The gonadal steroids probably act to inhibit localized inflammatory processes. Progesterone is generally regarded as immunosuppressive.

E. THE PITUITARY–THYROID AXIS: THYROTROPIN AND THYROID HORMONES

1. Regulation of Release

The major regulator of thyrotropin (i.e., TSH) secretion is the release of TRH from hypothalamic neurons located within the PVN. The TRH neurons receive input from norepinephrine neurons, which facilitate TRH release. The evidence in humans suggests that dopaminergic neurons and those containing somatostatin have an inhibitory influence on TSH release. AVP appears to facilitate TSH release. TSH release from thyrotrophs, in turn, stimulates the thyroid to release T_3 and T_4. These hormones exert negative feedback on the pituitary cells and may augment the activity of dopaminergic and somatostatin neurons, thereby providing further inhibitory feedback to TSH release.

2. Modulation of Immune Function

Less attention has been directed to the study of thyroid hormone effects on immune function than has been the case for the glucocorticoids or even the gonadal steroids. The available evidence has often been derived from conditions in which the thyroid is either overactive or underactive. In general, TSH and the thyroid hormones have been regarded as immunoenhancing. Excess of thyroid hormones can cause hypertrophy of lymph nodes and the spleen. T_3 receptors have been found on lymphocytes and the thymus. However, T_3 has been found to have nonlinear effects on the proportion of helper T/suppressor T cells in different immune compartments (e.g., spleen and peripheral circulation). Normal helper T/suppressor T ratios are observed only when T_3 has an intermediate (i.e., normal) level. Effects on delayed-type hypersensitivity and susceptibility to infection have appeared contradictory, because as thyroid hormones increase above a certain optimal level, they may trigger changes that serve to downregulate immune function. Abnormally high levels of T_3 and T_4 have been found to correlate with an increased level of soluble IL-2 receptor, a marker of T-cell activation, which actually may serve to block IL-2 binding to T cells and downregulate activation. NK cell activity also appears to be maintained by normal levels of T_3 and T_4. As these hormones decline with age, NK cell activity declines, a change that reverses if thyroid hormones are administered.

Thyroid hormones are also involved in the regulation of thymulin release by the thymus.

V. POSTERIOR PITUITARY GLAND HORMONES AND IMMUNE FUNCTION

A. ARGININE VASOPRESSIN

1. Regulation of Release

AVP is produced in neurons located in the supraoptic nucleus and the PVN of the hypothalamus. It is co-released with a specific neurophysin molecule. Neurons containing AVP project to sites other than the posterior pituitary. They are found to project to the median eminence and facilitate the release of ACTH. They also project to the forebrain and brainstem structures. In addition, AVP has been found in a variety of tissues throughout the body, including the thymus.

AVP release from the posterior pituitary is controlled by osmoreceptors located in anterior circumventricular structures (e.g., OVLT and SFO) and baro-receptors located in the heart and key blood vessels. These receptors project via afferent fibers in the vagus and the glossopharyngeal cranial nerves to the brainstem and on to the hypothalamus. Dopamine appears to stimulate AVP release, whereas norepinephrine can either stimulate or inhibit AVP release. Angiotensin II stimu-lates AVP release. Opioid peptides inhibit AVP release.

2. Modulation of Immune Function

Research on the immune effects of AVP is limited. AVP potentiates the action of CRH and, thus, can contribute to immune modulation via the pituitary–adrenal axis. It has also been found to have a direct effect on T-cell production of IFN-γ and, thus, appears capable of enhancing cell-mediated immune responses.

B. OXYTOCIN

1. Regulation of Release

Oxytocin release from the posterior pituitary originates in neurons located mostly within the supraoptic nucleus and the PVN of the hypothalamus. Oxytocin is not colocalized with AVP. Oxytocin neurons are found in other regions of the brain as well. Moreover, oxytocin is also produced by the gonads, the pineal gland, and the thymus gland.

Suckling provides a major stimulus for oxytocin release. Catecholaminergic and cholinergic neurons have been implicated in the release of oxytocin, but the effects are complicated. For instance, norepinephrine can increase or decrease oxytocin release by acting via α receptors or β receptors, respectively. Opioid peptides appear to inhibit oxytocin release.

2. Modulation of Immune Function

Research on the immune effects of oxytocin is quite limited. Oxytocin operates in a manner analogous to AVP and IL-2 to facilitate the production of IFN-γ by T cells. Thus, it can promote cell-mediated immune responses.

VI. OTHER HORMONES AND IMMUNE FUNCTION

A. INSULIN

1. Regulation of Release

Insulin secretion by the β cells of the islets of Langerhans is primarily regulated by the D-glucose level in the extracellular fluid bathing the β cells. Glucagon increases and somatostatin decreases insulin release via paracrine actions. Insulin release is stimulated by GH, cortisol, PRL, and the gonadal steroids. It is decreased by PTH. The effects of thyroid hormones are more variable. Epinephrine inhibits insulin release. Sympathetic nerve stimulation inhibits insulin release. Cholinergic stimulation promotes insulin release.

2. Modulation of Immune Function

It has been observed that as lymphocytes are becoming activated due to exposure to antigen or mitogen, they express insulin receptors. Insulin appears to facilitate T-cell growth in response to antigen. This seems more a matter of addressing the metabolic needs of the cell rather than prompting them to become activated or curtailing the process of activation. In essence, insulin facilitates the production of energy for the activation process.

B. PARATHYROID HORMONE

1. Regulation of Release

The release of PTH is predominantly regulated by the level of extracellular calcium. PTH is released when the calcium level decreases. The magnesium level has a similar but minor effect on PTH. β-Adrenergic stimulation increases PTH release, particularly when the calcium level is low. Histamine has also been observed, at least *in vitro*, to release PTH. PTH release is stimulated by calcitonin, glucagon, GH, cortisol, and prostaglandin E_2. A PTH-related peptide inhibits PTH release. Other inhibitory signals include somatostatin and prostaglandin E_1.

2. Modulation of Immune Function

Receptors for PTH have been found on leukocytes. PTH has been found to inhibit T-cell proliferation to mitogens and antibody production by B cells. The effect of PTH on B cells appears to be counteracted by IL-4. The impaired T-cell proliferation observed in cases of hyperparathyroidism can be corrected by parathyroidectomy.

C. MELATONIN

1. Regulation of Release

Melatonin is the major pineal hormone. Its synthesis and release are regulated by the light–dark cycle. The pineal senses the light conditions via a pathway originating in the retina, which projects to the suprachiasmatic nucleus and then to the PVN. The PVN gives rise to a pathway that ultimately ends up in the superior cervical ganglion. Postganglionic neurons innervate the pineal and release norepinephrine. The melatonin level is higher in the evening. Melatonin release is maximal during the early years of life and gradually declines with age.

2. Modulation of Immune Function

Melatonin appears to have immunoenhancing effects. It has been thought to have an oncostatic effect. The latter may be mediated via effects on immune responsiveness or through its function as a free-radical scavenger. Melatonin appears, in particular, to enhance antigen-driven T-cell–dependent humoral responses. However, melatonin's action does not appear to be direct and may be mediated by endogenous opioids. β-Endorphin can mimic the effects of melatonin. Melatonin induces activated T lymphocytes to produce opioid peptides, which, in turn, modify immune responsiveness. Melatonin has no effect on NK cell activity. Melatonin is ineffective in the absence of specific antigen. In general, there is an association between reduced melatonin and depressed immune function.

VII. THYMUS GLAND

Observations reported in the early 1900s indicated that removal of the thymus was associated with morphological changes in the adrenal cortex and the gonads. At that time, it was suspected that the thymus played an endocrine role. Over time, however, it has become clear that the thymus plays a central role in immune function. The importance of the thymus gland with respect to cellular immunity was discovered in the 1950s. Disorders such as DiGeorge syndrome demonstrated that when the thymus fails to develop, a lack of cellular immune function results.

A. PITUITARY REGULATION OF THYMUS

The thymic epithelium produces a variety of peptides that promote the maturation of thymocytes and appear to modulate neuroendocrine circuits. The release of these peptides, which include thymopoietin, thymulin, and thymosin, is influenced by various hormones. The existence of a specific thymus–stimulating hormone has been suspected but not demonstrated. The secretion of thymulin has been shown to be regulated by GH, PRL, cortisol, gonadal steroids, and thyroid hormones. In general, the maintenance of an appropriate endocrine environment is crucial for normal thymic function. For instance, even damage to the pineal induces thymic involution and causes defects in cell-mediated immunity.

The specificity of the hormonal influences on thymic function is supported by the discovery that thymic epithelial cells and thymocytes contain receptors for many hormones and peptides. However, because the thymic epithelium is able to secrete most of the hormones and peptides originally found in the pituitary, localized production may serve to regulate activity as well.

B. THYMUS REGULATION OF PITUITARY

Thymic peptides stimulate the release of GH, PRL, LH, and FSH. A thymosin component peptide, thymosin β_4, appears particularly effective in stimulating the pituitary–gonadal axis by promoting Gn–RH release at the level of the hypothalamus and LH release from the pituitary. Another thymosin component (MB–35 peptide) stimulates PRL and GH release. Thymopoietin analogues can enhance the production of POMC-derived peptides (e.g., ACTH and β-endorphin). Thymosin α_1 has been noted in some studies to have a downregulatory effect on TSH, ACTH, and PRL. Other studies, in contrast, have reported that thymosin α_1 stimulates cortisol release when injected into the cerebral ventricles.

C. HORMONES AND THYMOCYTE DEVELOPMENT

T-cell development and selection requires direct cell-to-cell interactions between thymocytes, thymic epithelial cells, and other cells that reside in the thymus (e.g., APCs). Cytokines play a key role in this process, which appears to be further supported by thymic peptides. As noted earlier in this chapter, production of both cytokines and thymic peptides is influenced by the hormonal composition of the microenvironment, which reflects both bloodborne levels and local release of hormones.

The hormonal modulation of activities within the thymus includes adhesion of thymocytes to the thymic epithelium, which is enhanced by T_3. Similar effects have been noted with PRL, GH, and IGF-1. Thymocyte proliferation can be stimulated by PRL. Effects on the composition of the T-cell repertoire have been observed as well. GH injected into aging mice increases the number of thymocytes. GH-deficient mice show low levels of thymulin and thymic hypoplasia and decreased $CD4^+$ and $CD8^+$ cells. Long-term treatment with GH restores thymic activity. In effect, hormonal influences within the thymus may affect the number of T cells produced, the makeup of the repertoire released with respect to $CD4^+$ and $CD8^+$ composition, and possibly even the diversity of antigen-recognition receptors.

VIII. CONCLUDING COMMENTS

This chapter has focused on endocrine influences on immune function, even though from the outset it was evident that endocrine and neural influences cannot be neatly separated (Figure 4.2). Thus, to speak of endocrine influences amounts to an oversimplification of what transpires, but it serves to highlight a class of molecules that participate in the communication between the "systems" described in this chapter. The fact that hormones once thought to originate within the pituitary are now known to be produced in a wide range of tissues further complicates the picture. However, it is important to remember that although most pituitary hormones can be secreted by various tissues, including the thymus, disruption of the pituitary supply has a major effect on immune function. In other words, the local effects of such hormones can rapidly fade if an organism is not able to sustain a systemic level, which in effect may act as a signal of the viability of the organism.

Hormonal influences appear as important background signals capable of sustaining and modulating the massive metabolic undertaking that constitutes a functional immune system. This sort of role seems most in line with the effects of insulin, GH, PRL, thyroid hormone, and PTH. Gender differences in immune function are evident and may be rooted in the fact that the female immune system

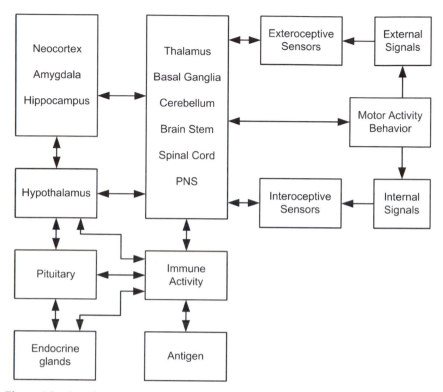

Figure 4.2   Central nervous system regulation of endocrine immunomodulation. PNS, peripheral nervous system.

must strike a balance between tolerance for some foreign antigen (e.g., sperm and fetus) on one side and intolerance for pathogen-derived foreign antigen on the other. This requires a modulation of immune functional characteristics in a manner sensitive to hormonal correlates of mating behavior, pregnancy, and gestation. The HPA axis appears to add yet another dimension of regulation, which takes into consideration important changes in context that could potentially threaten (i.e., stress) organismic well-being and may mobilize the immune system in ways that serve to optimize responses, at least in the short run.

In essence, hormonal influences on immune function may be classified into three major categories: (1) those that serve to coordinate the distribution of metabolic resources between the immune system and other life-sustaining tissues, (2) those that serve to modify immune activity vis-à-vis reproduction, and (3) those that serve to modulate immune responsiveness during periods of perceived potential threat caused by contextual change. These are important influences that integrate immune activation in response to antigen and cytokine costimulation into a more comprehensive set of survival functions.

IX. SOURCES

Ader, R., Felter, D., & Cohen, N. (eds) (1991), *Psychoneuroimmunology*, 2nd ed, San Diego, Calif, Academic Press.

Clark, R. (1997), "Somatogenic Hormone and Insulin Like Growth Factor-1: Stimulation of Lymphopoiesis and Immune Function," *Endocrine Reviews*, 18, pp. 157–179.

Conn, P.M. & Freeman, M.E. (eds) (2000), *Neuroendocrinology in Physiology and Medicine*, Totowa, NJ, Human Press.

Cutolo, M., Masi, A.T., Bijlsma, J.W.J., et al. (eds) (1999), "Neuroendocrine Immune Basis of the Rheumatic Disorders," *Annals of the New York Academy of Sciences*, 876.

Davis, S.L. (1998), "Environmental Modulation of the Immune System via the Endocrine System," *Domestic Animal Endocrinology*, 15, pp. 283–289.

DeGroot, L.J., Jameson, J.L., Burger, H., et al. (eds) (2001), *Endocrinology*, 4th ed, vols 1–3, Philadelphia, WB Saunders.

Jiang, Y., Yoshida, A., Ishioka, C., et al. (1992), "Parathyroid Hormone Inhibits Immunoglobin Production without Affecting Cell Growth in Human B-Cells," *Clinical Immunology and Immunopathology*, 65, pp. 286–293.

Klecha, A.J., Genaro, A.M., Lysionek, A.E., et al. (2000), "Experimental Evidence Pointing to the Bidirectional Interaction between the Immune System and the Thyroid Axis," *International Journal of Immunopharmacology*, 22, pp. 491–500.

McCann, S.M., Sternberg, E.M., Lipton, J.M., et al. (eds) (1998), "Neuroimmunomodulation: Molecular Aspects, Integrated Systems, and Clinical Advances," *Annals of the New York Academy of Sciences*, 840.

Tzanno-Martins, C., Futata, E., Jorgetti, V., & Duarte, A.J.S. (2000), "Restoration of Impaired T-Cell Proliferation after Parathyroidectomy in Hemodialysis Patients," *Nephron*, 84, pp. 224–227.

CHAPTER 5

Neuroimmune Modulation

I. Introduction   82
II. Peripheral Nervous System   82
   A. Somatosensory Pathways   83
   B. Visceral Sensory Pathways   84
   C. Autonomic Nervous System   84
   D. Enteric Nervous System   85
III. Peripheral Nervous System Innervation of Lymphoid Organs   86
   A. Bone Marrow and Thymus   86
   B. Spleen, Lymph Nodes, and Mucosa-Associated Lymphoid Tissue   86
IV. Chemical Signaling in the Periphery   87
   A. Classical Neurotransmitters   87
   B. Neuropeptides   88
   C. Cytokines   88
   D. Other mediators   89
V. Functional Effects of Peripheral–Neuroimmune Interactions   89
VI. Central Nervous System   92
VII. Bidirectional Central Nervous System–Immune System Interactions   94
   A. Subcortical Lesions   94
   B. Subcortical Responses to Immunization   95
   C. Neocortical Lesions   96
VIII. Learning and Immune Responses   98
IX. Personality and Immune Function   99
X. Concluding Comments   101
XI. Sources   102

I. INTRODUCTION

As described in Chapter 4, neural networks influence immune responsiveness via the endocrine system. Multiple neural pathways converge on the hypothalamic neurons that regulate pituitary hormone release and modulate their activity. These neural pathways convey both signals arising within bodily tissues (including the immune system) and signals initiated by events occurring in the individual's surroundings.

Neural effects on immune responsiveness are further mediated by the peripheral nervous system (PNS) innervation of the skin and the body's mucosal surfaces. Neural effects on immune function also occur as a result of direct innervation by the PNS of glandular tissue and the lymphoid organs. It is now well established that both primary and secondary lymphoid organs are innervated by the PNS. The juxtaposition of nerve terminals and leukocytes makes it likely that they interact in a functionally significant manner. Moreover, the interaction appears bidirectional. Neural signals modulate immune responses, and the latter modify neural activity. Immune effects on nerve terminals ultimately convey signals to the brain and may alter motivation, emotion, cognition, and behavior.

Indeed, the evidence suggests that immune cells may to some extent act in a sensory capacity, detecting the presence of pathogens or abnormal tissue conditions to mobilize protective responses. Such responses could be highly localized, occurring within a specific tissue's microenvironment or progressively enlisting higher order responses, including centrally mediated neuroendocrine changes capable of altering the emotional state and behavior of the individual. These same pathways can be traveled in the opposite direction and make it possible for brain activity underlying emotion and cognition to have an impact on immune responses as well.

II. PERIPHERAL NERVOUS SYSTEM

The PNS encompasses all neural elements outside of the brain and the spinal cord. It consists primarily of sensory and motor pathways carrying information to and from the tissues of the body. The sensory pathways communicate signals arising within visceral organs (visceral afferent pathways) and from the muscles, tendons, and skin. The motor pathways carry signals directed to the striated musculature and to the visceral organs. The visceral outflow is known as the *autonomic nervous system* (ANS) and consists of two principal subdivisions: the sympathetic and the parasympathetic branches. Another major component of the ANS is the enteric nervous system (ENS), which is contained within the walls of the gastrointestinal tract (GIT) (Figure 5.1).

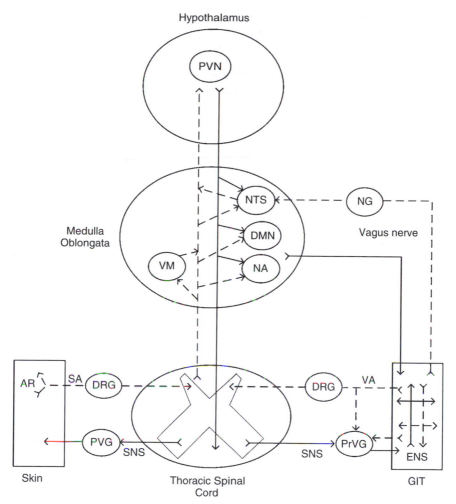

Figure 5.1 Peripheral nervous system regulation of immune activity. Major components include the enteric nervous system (ENS) located within the gastrointestinal tract (GIT), the parasympathetic nervous system (partially mediated via the vagus nerve), the sympathetic nervous system (SNS), and the somatic afferent (SA) and visceral afferent (VA) fibers. AR, axon reflex; DMN, dorsal motor nucleus of X; DRG, dorsal root ganglion; NA, nucleus ambiguus; NG, nodose ganglion; NTS, nuclei tractus solitarii; PrVG, prevertebral ganglion; PVG, paravertebral ganglion; PVN, paraventricular nucleus; VM, ventral medulla.

A. SOMATOSENSORY PATHWAYS

The skin is covered with sensory receptors that respond to touch, temperature, and pain stimuli. The pain receptors (nociceptors) are also sensitive to chemical agents released in their vicinity. Such chemical agents are released as a result of tissue injury, which serves to activate resident (intraepithelial) immune cells. As a consequence, the nerve endings release peptides capable of influencing local immune activation and transmit messages to the central nervous system (CNS).

The somatosensory axons that innervate the body's skin have cell bodies located within the dorsal root ganglia (DRG) and project to neurons within the spinal cord on the same side (ipsilaterally) as the DRG. The spinal interneurons then project to motor neurons in the region and cross the midline and travel up toward the brainstem as part of ascending projections in the anterolateral region of the spinal cord. Many of these axons terminate within the reticular formation (brainstem), the periaqueductal (midbrain) gray matter, and the dorsal thalamus. These neural structures have been implicated in the regulation of arousal and the processing of pain signals. Input from the facial skin is similarly organized, except that communication with the CNS is accomplished via the fifth cranial nerve (trigeminal). The ultimate effects of the centrally transmitted signals arising from immune activity have not been fully traced but can be expected to affect somatic reflexes, autonomic activity, endocrine function, and behavior.

B. VISCERAL SENSORY PATHWAYS

Visceral sensory (afferent) neurons convey signals from internal organs and mucosal surfaces to the CNS. These neurons have cell bodies located in the DRG and ganglia associated with cranial nerves. Historically, these pathways have been viewed as independent of the ANS. However, given their close association to the regulation of ANS activity, some investigators have argued for their inclusion as sensory channels for the ANS, particularly because visceral afferents are typically found in the same nerves (e.g., vagus nerve and splanchnic nerves) that carry the autonomic outflow to the viscera.

Like the somatosensory nociceptive afferents, when stimulated the visceral afferent neurons release peptides in the tissues they innervate. Moreover, they influence autonomic outflow both in the periphery at the postganglionic level and centrally within the spinal cord. Many visceral afferent fibers are normally inactive and become responsive only under conditions of tissue injury or that threaten such injury. The activity appears capable of modifying regional responses and autonomic outflow, and again though difficult to trace would be expected to cause CNS changes capable of affecting the psychological state of the individual.

C. AUTONOMIC NERVOUS SYSTEM

The ANS is that part of the PNS outflow that does not innervate the striated musculature. Instead it serves as the major neural input to visceral organs, the vasculature, and glandular tissues. It is composed of the sympathetic and parasympathetic branches. The ENS has also been included as part of the ANS, even though it is entirely contained within the GIT.

The parasympathetic and sympathetic branches are primarily concerned with conservation or mobilization of energy, respectively. They can be differentiated on the basis of neuroanatomical considerations. The sympathetic branch originates from neurons located in the thoracic and lumbar regions of the spinal cord. The parasympathetic branch projects via cranial nerves III, VII, IX, and X, exiting from the brainstem and neurons located in the sacral region of the spinal cord.

The neurons of the ANS with cell bodies in the brainstem and spinal cord are known as *preganglionic autonomic neurons*. They do not make direct contact with the tissues they innervate. They project to ganglia (collections of neuron cell bodies) in the periphery, which are located either at some distance from (sympathetic), very near, or sometimes inside (parasympathetic) the target organ. The preganglionic neurons synapse on the neurons located within the ganglia. The latter give rise to postganglionic neurons that project to the target tissues. The ganglia are arranged along side of the spinal column (paravertebral ganglia) or are found more distally (prevertebral ganglia). These ganglia receive input not only from the preganglionic neurons, but also from the visceral afferent neurons and neurons of the ENS. Thus, the postganglionic signals are subject to multiple influences and are not simply a reflection of the preganglionic message.

The sympathetic and parasympathetic branches of the ANS have also been differentiated based on the predominant neurotransmitter released by the postganglionic neurons. Sympathetic pathways predominantly employ norepinephrine (NE), except in the case of axons innervating the sweat glands, which release acetylcholine. The parasympathetic pathways employ acetylcholine. Preganglionic neurons of both branches use acetylcholine. In addition, postganglionic neurons contain peptides that appear to be co-released with the classical neurotransmitters (i.e., NE and acetylcholine). Sympathetic terminals have been shown to contain neuropeptide-Y (NPY), somatostatin (SS), and galanin (GAL). Parasympathetic terminals contain vasoactive intestinal peptide (VIP), substance P (SP), and calcitonin gene-related peptide (CGRP). The presence of peptides appears less evident in preganglionic neurons, but they have been found to contain enkephalins and corticotropin-releasing hormone. Moreover, to complicate matters, ANS terminals are apparently capable of switching their profile of neurotransmitter/neuropeptide release based on local signals and the frequency of stimulation. Evidently, these terminals are dynamically controlled and can adapt to local conditions.

D. Enteric Nervous System

The ENS resides entirely within the walls of the GIT. It receives input from both the sympathetic and the parasympathetic branch of the ANS. However, the input is sparse relative to the rich interconnectivity of the enteric neurons, which have been referred to as the *second brain*. These neurons, which are as numerous as those found in the spinal cord, are involved in processing local signals and give rise to both motor (GIT contractions) and secretory responses (release of hormones and enzymes) by the gut. The cell bodies are grouped in small clusters (i.e., enteric ganglia). They interconnect to form two plexuses, the myenteric plexus (evident throughout the entire GIT) and the submucosal plexus (present only in the intestines). The enteric neurons contain a wide range of neurotransmitters and neuropeptides.

In addition to its own nervous system, the gut contains its own population of endocrine cells scattered among the epithelial cells. These cells release numerous peptides capable of influencing the nerve cells in the vicinity of and in the gut-associated lymphoid tissue. The gut appears as one of the most critical interfaces between the organism and antigen arising from the outside world. A precarious balance must be struck between underreacting and overreacting to the presence of antigen. In fact, because of the overwhelming presence of antigen, the gut remains in a constant state of subclinical inflammation.

III. PERIPHERAL NERVOUS SYSTEM INNERVATION OF LYMPHOID ORGANS

A. BONE MARROW AND THYMUS

Sympathetic innervation of both the bone marrow and the thymus has been demonstrated. Postganglionic neurons send axons along the blood vessels that enter the bone marrow and thymus. They then venture into areas populated by immune cells undergoing development. Parasympathetic innervation has been more difficult to demonstrate, but it appears that at least in the thymus, such innervations may be present. Again, the nerve terminals appear to intermingle with thymocytes. Visceral afferent nerve fibers with cell bodies in the DRG and cranial nerve ganglia are also known to innervate the primary lymphoid organs.

B. SPLEEN, LYMPH NODES, AND MUCOSA-ASSOCIATED LYMPHOID TISSUE

The spleen receives sympathetic innervation arising from the mesenteric/coeliac ganglia. The fibers approach the spleen with the splenic nerve, enter at the pilar region, and distribute themselves along the blood vessels, innervating them and the smooth muscle cells, and extend into areas populated by immune cells. Parasympathetic innervation remains controversial because even though there is evidence of cholinergic substances in the spleen, they do not appear to be localized within neural elements for the most part. Visceral afferent nerve fibers are also present within the spleen. Innervation of the lymph nodes is generally similar to that of the spleen.

Mucosa-associated lymphoid tissue is innervated via the sympathetic, parasympathetic, and visceral afferent systems, and, in the case of the gut, the ENS. Endogenous neural elements (microganglia) are found in other visceral organs (e.g., the lungs) but are not as extensive as what is found in the gut.

IV. CHEMICAL SIGNALING IN THE PERIPHERY

The juxtaposition of neural processes and immune cells in various tissues is not sufficient to demonstrate functional interaction, although it clearly increases the probability that such interaction could occur. Investigators have labored to demonstrate that the nerve fibers are not simply present to modulate regional blood flow or smooth muscle activity. This seems very much like a concern rooted in the vestigial notion that immune responses must be largely autonomous. The likely scenario is that the neural terminals within a microenvironment have multiple effects on the cellular elements in the vicinity and that what is most critical is the presence of receptors in or on the cells capable of binding and responding to substances released by the nerve terminals.

Chemical communication has been found to be multidirectional, forming local circuits that are capable of dynamic change over time. For instance, there can be changes in the profile of receptors expressed by the cells in the area, the types of substances released, and/or the coupling to intracellular responses, which ultimately alter gene expression within the cells. In essence, interactions defy simple characterization. Nonetheless, the major molecular candidates for such communication have begun to be identified and include the classical neurotransmitters and neuropeptides on the neural side and the cytokines on the immune side.

A. CLASSICAL NEUROTRANSMITTERS

Leukocytes have been shown to possess specific receptors for the classical neurotransmitters of the ANS. Leukocytes express α- and β-adrenergic receptors and muscarinic cholinergic receptors. Serotoninergic, 5-HT_{1A} receptors, and dopamine receptors have also been found. It should be noted that immune cells appear capable of synthesizing the classical neurotransmitters, so, as was the case with hormones, autocrine and paracrine interactions independent of neural activity are a possibility.

B. NEUROPEPTIDES

A long list of peptides have been found to modulate neural activity and are often co-released along with the classical neurotransmitters. Evidence has also been obtained that immune cells express receptors for many of the neuropeptides, and as a result, such peptides would be able to influence immune activity.

The list of peptides for which receptors have been found on immune cells include SP, CGRP, NPY, SS, VIP, β-endorphin, enkephalins, GAL, gastrin-releasing peptide, peptide histidine isoleucine amide (PHI), pituitary adenylate cyclase–activating polypeptide, and others. As previously noted, these peptides are colocalized with the classical neurotransmitters in the ANS and are contained within the visceral afferent terminals. However, the aforementioned peptides are not exclusively expressed in nerve terminals, so more complex interactions within specific microenvironments are possible.

C. CYTOKINES

Local circuits whereby immune cells are able to communicate back to nerve terminals involve the cytokines released in the process of immune activation within microenvironments. There is compelling evidence that immune responses in the periphery have an impact on CNS activity. One avenue of communication is thought to be the bloodstream, although a number of difficulties are immediately apparent, such as the magnitude of the signal and the loss of spatial information. Nonetheless, substances released by the immune system can enter the circulation and through effects on areas of the CNS that lie outside the blood–brain barrier (circumventricular organs) can give rise to a cascade of effects that ultimately alter neural activity. There are a number of circumventricular structures through which such interactions are possible. They include the organum vasculosum of the lamina terminalis (OVLT), the subfornical organ (SFO), and the area postrema (AP).

A second avenue of communication entails local activation of nerve terminals that give rise to local responses (axon reflexes) and to signals that travel into the CNS. A cytokine that has been extensively studied in this regard is interleukin-1 (IL-1). The axon reflex component can cause local vascular permeability changes and promote the inflammatory response. IL-1 is also known to participate in the induction of systemic changes such as increased body temperature (i.e., fever), a physiological change that requires central mediation and serves to fight pathogens.

The IL-1 signal appears to be transmitted primarily by visceral afferent fibers carried within the vagus nerve. The neural signal may be directly initiated either by IL-1 because there is some evidence that the visceral afferent nerve terminals may express IL-1 receptors or through the microenvironmental effects of IL-1.

D. OTHER MEDIATORS

Substances other than those previously described are likely involved in the localized signaling and may give rise to centrally transmitted activity. Substances such as the proinflammatory mediators released by mast cells, as well as gaseous molecules such as nitric oxide (NO), appear to be likely candidates.

V. FUNCTIONAL EFFECTS OF PERIPHERAL NEUROIMMUNE INTERACTIONS

The functional effects of neuroimmune interactions, as the term implies, should depend on where in the body they take place and at what point in time during the development of the immune response. Moreover, they should depend on leukocyte type, the specific substances released by nerve terminals in the region, and the background of chemical stimulation occurring from other sources at the moment within the microenvironment where the cells under scrutiny are inter-acting.

Major aspects of the immune process that appear susceptible to neural modulation include the following: (1) the generation and differentiation of immune cells within the primary lymphoid organs, (2) innate immune responses such as phagocytosis, antigen presentation, mast cell degranulation, and natural killer (NK) cell activity, (3) adaptive immune responses such as B-cell proliferation, antibody production, T-cell proliferation, cytokine synthesis and release, and T-cell cytotoxicity, and (4) leukocyte traffic, a critical aspect of immune activity that permits encounters with antigen, activation of immune cells, and homing to sites of infection or tissue abnormality.

The current state of knowledge does not yet allow detailed characterization of the role of any specific neurotransmitter on the function of the immune system or even on some circumscribed aspect of its function. This can be illustrated by looking at the picture that has emerged with respect to the role of NE in modulating immune responses. NE is probably the most extensively studied of the classical neurotransmitters from the perspective of neuroimmune modulation. NE is primarily released by sympathetic nerve terminals, although it originates from the adrenal medulla as well, and it may even be released by leukocytes. NE can downregulate its own release by its action on presynaptic α_2-adrenergic receptors and, thereby, modulate its availability to other cells. Moreover, sympathetic nerve terminals release NE in conjunction with other neuromodulators such as NPY. Therefore, observed effects *in vivo* cannot be simply attributed to the release of NE.

β-Adrenergic receptors have been found on lymphocytes, with the density of receptors varying across cell types (e.g., B cells express more receptors than T cells). β-Adrenergic receptors have also been found on other immune cells (e.g., dendritic cells), and β-adrenergic receptors are subject to downregulation in the face of sustained stimulation. Thus, the influence of NE on other cells can in principle be held constant, by downregulation either of NE release or of the expression of its receptors on other cells.

The SNS innervates the bone marrow and the thymus; therefore, in all likelihood NE has the opportunity to modulate the proliferation and differentiation processes that take place in those organs. Indeed, some evidence suggests that cell proliferation in the bone marrow can be increased by destruction of the NE terminals. The chemical destruction of NE terminals by the neurotoxin 6-OHDA has been shown to cause increased antibody production to specific antigen but in a strain-dependent manner. Evidently, genetic factors influence the magnitude of the effect of NE denervation. Delayed-type hypersensitivity (DTH), a cell-mediated response, is diminished by sympathetic denervation of an area's draining lymph node. However, caution is necessary in interpreting this observation because denervation of a region can alter leukocyte traffic into the lymphoid tissue.

SNS activation has been consistently found to cause a rapid and transient leukocytosis (i.e., increased white blood cell count in peripheral circulation). In particular, NK cell number increases in the peripheral circulation but decreases in the spleen. SNS activation generally inhibits T-cell function and NK cell activity *in vitro*. It downregulates proinflammatory cytokines such as tumor necrosis factor-α (TNF-α). β-Adrenergic stimulation increases plasma IL-10 while decreasing IL-12 and interferon-γ (IFN-γ), a cytokine profile that does not favor Th1 responses. Evidently, both NE denervation and activation of NE receptors can cause similar effects with respect to cell-mediated immunity. In both cases it appears to be diminished, so it is not possible to assign a simple linear effect to the impact of NE activity on immune responsiveness. Clearly, despite sustained effort to characterize the SNS influence on immune responsiveness, many questions remain unanswered and the picture promises to be complex.

The role of peptides on immune responsiveness has been the subject of investigation, and it is not entirely separable from that of classical neurotransmitters because they are often co-released. Although peptides have multiple sources, nerve terminals are a major source of peptides within any given microenvironment. They are released from visceral afferent fibers and from the autonomic innervation of the body. Receptors for peptides such as SP, CGRP, NPY, VIP, and SS, among others, have been found on leukocytes. As a rule, neuropeptide receptors are more numerous in tissue lymphocytes than on those in the circulation. This seems consistent with the role of peptides in the regulation of highly localized responses.

Among the more extensively studied peptides are SP, CGRP, NPY, VIP, and SS. SP promotes inflammatory reactions including those that appear to be neurogenic. It can induce the synthesis of mediators of inflammation that are not normally stored within cells (e.g., leukotrienes). It stimulates the production of IL-1, IL-6, and TNF-α. SP stimulates monocyte chemotaxis and enhances T-cell proliferation and antibody production, particularly immunoglobulin A (IgA), by B cells. SP also enhances NK cell activity, at least in some tissues. Overall, the most consistent finding with respect to SP is its facilitation of inflammation.

CGRP inhibits the maturation of B cells. It interferes with the ability of dendritic cells to present antigen and inhibits IL-1 production. It diminishes DTH and decreases T-cell proliferation in response to antigens, probably by down-regulating IL-2 production. CGRP appears to influence Th1 responses more than those promoted by Th2. Thus, its downregulating effect would be most evident in the case of cell-mediated immunity. CGRP has also been found to enhance the chemotaxis of neutrophils and eosinophils to sites of infection. SP and CGRP are often colocalized in the PNS, so the potential exists for their effects in some respect to antagonize each other and perhaps even cancel out, depending on relative quantity released and receptor expression in target cells.

VIP may inhibit the maturation of immune cells in bone marrow by counteracting the effect of IL-7. In the thymus, VIP may protect thymocytes from signals that initiate apoptosis. It can inhibit production of IL-2 and IL-4, as well as inhibiting NK cell activity. VIP can enhance IgA and IgE production. It can facilitate monocyte migration and histamine release from mast cells and block lymphocyte traffic (i.e., keep lymphocytes at their current locations—in or out of the secondary lymphoid tissues).

NPY has been found to have a generally inhibitory effect on antibody production to T-cell–dependent antigens. It inhibits NK cell activity and can increase lymphocyte proliferation. Through actions on endothelial cells, NPY can promote the movement of leukocytes into infected tissues.

SS has an inhibitory effect on lymphocyte proliferation. It decreases antibody synthesis, NK cell activity, and mast cell degranulation. SS receptors are more evident on activated lymphocytes; thus, its effect may come into play after cells have begun to respond.

In addition to the classical neurotransmitters and the neuropeptides, there are modulators such as NO, which is produced by a variety of cells including those that line the blood vessels, visceral organs, and cells of the immune system and neurons. NO release can serve to modulate cell activities and has antimicrobial properties. It can also suppress antibody production and dampen lymphocyte proliferation.

The preceding overview should drive home the point that only the surface has been scratched of what are complex spatiotemporal processes. So far, most observations are the result of *in vitro* experiments that have not addressed realistic scenarios of neuropeptide/neurotransmitter release. Therefore, realistic models of what may transpire *in vivo* are not available. The trends that have emerged suggest that there are multiple inhibitory and excitatory modulators that may have only partially overlapping actions. Consequently, the responsiveness of the immune cells would be expected to be generally dampened until there is a shift favoring one class of modulator (i.e., inhibitory or excitatory).

VI. CENTRAL NERVOUS SYSTEM

All structures located within the spinal cord and the brain make up the CNS. In this context the focus is on major regions of the CNS that at least based on neuroanatomical evidence (i.e., connectivity) appear integral in neural networks capable of regulating neuroimmune interactions. For instance, studies have shown that if the spleen is infected with a virus that is taken up by nerve terminals and travels into the CNS through neurons that make synaptic contact (i.e., a neurotrophic virus), it is eventually found all the way up to the paraventricular nucleus (PVN) of the hypothalamus. Conversely, if a chemical that is transported from the cell body to the neuron's terminals (i.e., anterograde tracer) is injected into the PVN of the hypothalamus, it is later detected in the intermediolateral region of the spinal cord, where the preganglionic sympathetic neurons are located.

Neural networks regulating immune function obviously include the neurons that give rise to the sympathetic and parasympathetic outflow located within the spinal cord and the brainstem. These neurons are located primarily within the dorsal motor nucleus of the vagus, the nucleus ambiguus, and the intermediolateral column of the spinal cord. These networks also incorporate neurons that are part of cutaneous and visceral sensory ascending projections. The neurons give off collaterals or terminate in brainstem structures such as the nuclei tractus solitarii (NTS), nuclei in the ventral medulla, the parabrachial nuclei, and the periaqueductal gray. These structures, in turn, are interconnected and project back to the neurons giving rise to the autonomic outflow, as well as those mediating sensory input from the periphery.

There is evidence that the thalamus and the hypothalamus are incorporated in CNS neuroimmune networks. Specifically in the thalamus, the ventral posterior and midline nuclei are a way station for visceral sensory input, and within the hypothalamus the PVN sits at a crossroad for the integration of visceral, endocrine, and other sensory signals that serve to regulate growth-promoting and life-sustaining activities.

The hypothalamus, the thalamus, and the brainstem structures implicated in neuroimmune network are subject to modulation by higher order neural systems, which are generally regarded as substrates for motivation, emotion, cognition, and

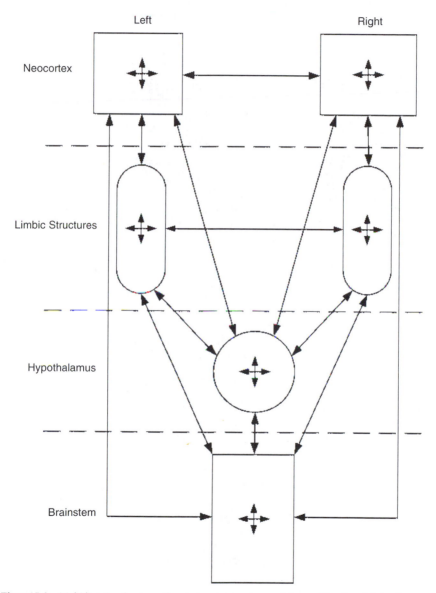

Figure 5.2   Multidirectional nature of central nervous system activity capable of modulating immune activity.

the conscious awareness of such processes. Included among the higher order structures are the amygdala, the hippocampus (components of the classical limbic system), and much of the neocortex, particularly the insular cortex, the anterior cingulate cortex, and the ventromedial prefrontal cortex. There are extensive interconnections between the amygdala, the bed nucleus of stria terminalis (BNST), and the hypothalamus. The hippocampus is also extensively connected to the hypothalamus. Both the amygdala and the hippocampus receive major inputs from the multimodal neocortex. Indeed, it is fair to assume that most if not all areas of the brain are able to influence each other at least indirectly (Figure 5.2). Thus, in effect, there are well-established links between higher order brain systems and the neurons known to make direct contact with immune cells. Such links would permit the brain functions that constitute psychological processes to modulate the activity of the immune system. In turn, immune activity would be able to have an impact on psychological processes, both through influences mediated by visceral/cutaneous afferent pathways and via circumventricular organs (e.g., OVLT, SFO, and AP), which are sensitive to macromolecular substances in the circulation (e.g., cytokines) and project to the brainstem and hypothalamic nuclei that are integral to neuroimmune networks.

VII. BIDIRECTIONAL CENTRAL NERVOUS SYSTEM–IMMUNE SYSTEM INTERACTIONS

Evidence has accumulated over the last 20 years demonstrating that lesions of the CNS are associated with changes in immune function. These studies have not been typically concerned with the concomitant behavioral and emotional changes produced by the lesions. However, when considered in light of other studies that have examined behavior, it is possible to make inferences regarding how specific structures (e.g., hippocampus) may play a role in modulating behavior and immune function. Evidence suggesting functional overlap is crucial to the notion that aspects of brain activity (i.e., psychological processes) can modify immune responses. This line of evidence is bolstered by research showing that the initiation of an immune response can have measurable effects (e.g., changes in electrical activity or alterations in neurotransmitter levels) within the CNS and modify the behavior of the organism (e.g., induce sickness behavior), further underscoring the bidirectional nature of neuroimmune interactions.

A. SUBCORTICAL LESIONS

Lesions in the caudal reticular formation, which contain many of the NE neuron nuclei, have been associated with diminished DTH responses, whereas more rostral reticular formation lesions, in the area of the 5-HT–containing raphe nuclei, lead to enhanced DTH responses. Damage to the nearby parabrachial nuclei is associated with a decrease in the proliferative response by thymocytes. More rostral lesions located in the substantia nigra, one source of ascending dopaminergic neurons, produce immune changes that depend on the side of the lesion. Left-sided lesions produce decreases in immune responsiveness, whereas right-sided lesions have no effect or lead to enhanced immune responsiveness. The substantia nigra influence is similar to what is observed after neocortical ablations (see the section "Neocortical Lesions," later in this chapter). Still more rostrally, damage to the nucleus basalis, which is the source of significant cholinergic innervation to the neocortex, produces increased lymphocyte responsiveness to mitogens and enhanced NK cell activity.

Lesions of the anterior hypothalamic area are associated with a decrease in the number of splenocytes (leukocytes in the spleen) and thymocytes (immature T cells in thymus). Functional studies have found decreased NK cell activity, decreased antibody production, and diminished DTH, an index of cell-mediated immunity. Some of these effects could be reversed by hypophysectomy, noteworthy because hypophysectomy has generally been found to be immunosuppressive, at least when the hypothalamus remains intact. Evidently, damage to the anterior hypothalamus releases downregulatory signals to the immune system that are reversible by subsequent damage to the pituitary. The effect of anterior hypothalamic lesions is transient, even if the pituitary is left intact. Thus, the immune system is able to reequilibrate spontaneously. It is likely that hypophysectomy hastens the reequilibration process but may eventually lead to a more sustained state of immune suppression.

Moving higher up the CNS, lesions to structures such as the septal region, the amygdala, the BNST, and the hippocampus affect immune responsiveness. These structures are typically included as part of the limbic system, an interconnected set of structures implicated in the regulation of emotional responses and the formation of some forms of memory. The involvement of these structures in immune regulation brings us much closer to the notion that emotional state can modulate immune responses and vice versa. Lesions to both the amygdala and the hippocampus produce transient increases in the number of splenocytes and thymocytes, as well as the augmentation of mitogen-induced lymphocyte proliferation. Hypophysectomy can reverse the effect of lesions to the amygdala and hippocampus. Septal and BNST lesions produce a transient decrease in NK cell activity.

B. Subcortical Responses to Immunization

Changes in the electrical activity of neurons within specific CNS structures have been observed after inoculation with specific antigens. For instance, increased neuronal firing has been observed within the ventromedial nucleus of the hypo-thalamus on the day of peak antibody production to a novel antigen. Nonlinear changes in activity have also been found for some nuclei (e.g., PVN), in which inoculation is initially associated with a decrease in neural activity below baseline, followed by a gradual increase over a period of 6 days.

Inoculation with a novel antigen has been observed to decrease NE within the hypothalamus by 4 days after inoculation. NE reduction appears highly specific to the PVN and coincides with maximal antibody production occurring on day 4 and increased glucocorticoid levels in the circulation. Evidence of increased NE release is evident within 30–90 minutes after inoculation with antigen. Thus, the gradual changes in neurotransmitter levels may reflect disparities between utiliza-tion and the neurotransmitters rate of synthesis or transport into the specific region. Changes in neurotransmitter levels are not solely restricted to the hypothalamus. They have also been documented to occur within the NTS and the hippocampus. In addition to changes in the levels and release of classical neurotransmitters contained within the ascending brainstem projections, there is evidence that induction of immune responses in the periphery causes the release of cytokines (e.g., IL-1) within the CNS. Cytokine release within the CNS is thought to give rise to metabolic and behavioral changes, collectively known as the *sickness response*.

C. Neocortical Lesions

The ablation of neocortex has been studied with respect to the effect on immune responsiveness. The findings have shown that the effect depends on which hemisphere is ablated. Lesions to the left hemisphere have been shown to be primarily associated with decreased T-cell numbers, lymphocyte proliferation, and NK cell activity. In contrast, right hemisphere lesions produce the opposite effect with respect to T-cell responses but do not affect NK cell activity. However, when both hemispheres were lesioned, the effect was like that of left hemisphere lesions. These findings are especially noteworthy because evidence suggests that the hemispheres function asymmetrically with respect to various cognitive func-tions and emotional responses. In particular, the right hemisphere appears intim-ately involved in the experience of negative affect. Thus, the observation that damage to the right hemisphere is associated with immune enhancement appears in line with the notion that negative emotions are immunosuppressive. However, the research showing that neocortical lesions have asymmetrical effects on immune function has been done using laboratory animals. Whether humans are similarly affected can be addressed only by examining immune function in patients with lateralized brain damage.

Studies of patients who have experienced brain trauma or stroke or have been diagnosed with a brain tumor all show evidence of inflammatory responses within the brain. Proinflammatory cytokines are elevated within the cerebrospinal fluid (CSF) and activation of microglia and astrocytes is evident, as is upregulation of cell adhesion molecules in vascular endothelium in the region. The latter permits movement of leukocytes into the region. A similar picture has been observed in the case of neurodegenerative processes associated with dementia. Intracerebral immune activation can be part of what causes neural damage or, as is suggested by some of the evidence, may be protective by upregulating production of nerve growth factor (NGF) and downregulating apoptosis in the region, as is true in the periphery when inflammation promotes destruction of pathogen and activates tissue repair processes.

Injury to the brain appears to be associated with diminished cell-mediated immunity, an effect that may explain the increased susceptibility to infections that is frequently observed. This is particularly the case during the acute phase of the lesion. As time progresses, individuals who survive show evidence of enhanced immune responsiveness as indexed by DTH. DTH appears more pronounced on the side of the body that is affected (paretic side). The DTH response is positively correlated with axon-reflex–mediated vasodilation in the region. This is most evident when the lesion is located on the right side of the brain and affects the frontal cortex and the putamen. DTH asymmetry is not as evident when lesions are bilateral or are primarily found on the left hemisphere. Indeed, lesions that are relatively large and located on the right hemisphere frontally appear to enhance the DTH response overall. Other studies have shown that intracranial tumors located within the left hemisphere are associated with a decrease in lymphocyte proliferation. Right-sided tumors do not appear to suppress lymphocyte proliferation. These observations of asymmetry in humans appear at least superficially consistent with the animal research suggesting lateralization of cortical regulation of immune responsiveness. The evidence is indicative of the possibility that right hemisphere overactivity tends to be immunosuppressive, which is what would be expected if right hemisphere overactivity would also contribute to negative affective experience.

It should be noted in this context that individuals who show evidence of being right hemisphere dominant, by being either left handed or afflicted with language disabilities, have been reported to be more frequently affected by immune-mediated disorders, specifically autoimmune and allergic disorders. Such disorders are indicative of immune dysregulation, rather than immune suppression. However, one way to integrate these observations is to postulate that if the right hemisphere must subserve multiple functions including handedness and speech, which are typically based on the left side, it may be less able to modulate immune responsiveness. Essentially, what these findings establish is that neural networks incorporating neocortical circuits are capable of modulating immune responses and that structures as high as the limbic system show changes related to the initiation of immune responses in the periphery. These same structures participate in the elaboration of emotional responses to environmental events and are integral to the learning process necessary for adaptive function within specific contexts. Thus, it seems likely that emotional states and cognitive processes will affect immune function and that immune activity must have effects on emotion and cognition.

VIII. LEARNING AND IMMUNE RESPONSES

Some of the earliest efforts to demonstrate neural regulation of immune function examined the learning process. Investigators sought to condition immune responses to neutral stimuli, which do not naturally elicit immune responses. This form of learning, called *classical conditioning*, was discovered by Ivan Pavlov. However, it was not until Robert Ader and his colleagues began to publish methodologically rigorous studies that this aspect of neuroimmune regulation began to be more generally accepted.

A substantial literature has accumulated over the past 25 years. The studies do not always permit strong conclusions regarding "learning" but overall make it highly likely that a neutral stimulus ("conditioned stimulus," or CS), by being temporally associated with an immune-modulating stimulus ("unconditioned stimulus," or UCS), can come to modulate immune responsiveness through a process that is thought to be dependent on CNS mediation. Studies have shown that if a neutral stimulus (i.e., a CS), such as a taste, a smell, or a more complex multisensory event that does not directly affect immune function, is paired on at least one or a few occasions with a stimulus that naturally elicits an immune response (i.e., produces an "unconditioned response" [UCR]), then the neutral stimulus gains the property of eliciting a response similar to that produced by the natural immune-modulating stimulus.

Studies conducted by Ader and other investigators have provided evidence that the effect of an immunosuppressive drug (UCS) can be elicited by the taste of saccharin (CS) after conditioning. It is also possible to evoke the immunosuppressive effect of stress (UCR) by situational cues (CS) that are not intrinsically noxious but are present during noxious stimulation. Research has shown that immune enhancement, elicited by drugs, cytokines, or specific antigen, is conditionable as well. However, it is often necessary to present both the neutral stimulus (CS) and a suboptimal dose of the natural activator (UCS) to see the effect of the conditioning process. In other words, the conditioning process creates an underlying state whereby the neutral stimulus (CS) potentiates the ability of the natural stimulus (UCS) to elicit changes in immune activity.

Some of the studies have been conducted with human volunteers. These studies underscore that learning via the conditioning process serves to establish expectations or creates anticipation of what is to come. Here, one also encounters a bit of a dilemma in that some of the evidence indicates that the learning process can lead to immune suppression, which is not generally thought to be adaptive. Therefore, one is left with the possibility that either what is learned in some circumstances may not be adaptive or that our assumption that immunosuppression is generally maladaptive may be incorrect.

The importance of this work cannot be overstated. It makes it clear that a variety of stimuli can access the immune system via the nervous system. The specific pathways involved remain to be established. They could be entirely peripheral (e.g., signals from the gut acting through paravertebral ganglia could affect a variety of immune tissues). More likely, the CNS is involved, and this is supported by preliminary work showing that lesions to structures in the limbic system (amygdala) and the neocortex (insular cortex) interfere with the conditioning of immune responses. Amygdala lesions interfere with the acquisition but not the expression of a previously acquired association. Insular cortex lesions block expression and presumably acquisition of an association between the neutral and some immunomodulating stimulus.

This work also has underscored the existence of marked individual differences in the extent to which organisms learn or become conditioned. Some do not seem to form associations as readily as others under similar experimental conditions. This may be the level at which adaptiveness enters the picture; in the sense that without prior knowledge of whether an association is going to improve survival, it would seem best if organisms were dispersed with respect to forming specific associations.

IX. PERSONALITY AND IMMUNE FUNCTION

The notion that there is a connection between personality and disease goes back to antiquity. In more modern times, personality characteristics were often thought to be associated with psychosomatic disorders. Most recently, the role of personality as a risk factor in various diseases has come to the forefront. Certainly, much has been written about personality in relation to cardiovascular disorders. Now, investigators are actively exploring the role of personality on immune responsiveness. Interest in personality with respect to disease has been a natural outgrowth of the fact that there are individual differences in the response to disease-causing conditions or organisms.

Personality is one of those psychological terms that is part of everyday language. However, it is important to be clear that more than anything else, it implies stability; that is, an individual's personality comprises behavioral expressions, emotional reactions, and thought patterns that recur over time and across situations. This relative invariance is fundamentally a manifestation of dynamic aspects of brain organization that tend toward perpetuity. Such aspects of brain organization emerge as a result of genetic factors being acted on by environmental forces and leading to the sculpting of neural networks in ways that constrain their response characteristics. Loss of plasticity gives rise to the stability of personality, and this can occur quite early, even prenatally. Characteristics that appear early in life are subsumed under the rubric of temperament, whereas those that are evident by late adolescence are regarded as facets of personality. Again, when one speaks of temperament or personality, one is referring to dynamic aspects of brain organization. For instance, Davidson *et al.* (1999) have shown that individuals who experience sustained negative affect tend to show an asymmetrical pattern of activation over the frontal regions of the scalp. The right side appears more activated relative to the left, as indexed using recordings of electrical activity. This pattern has been observed throughout the life span, beginning in infancy. Thus, it must reflect aspects of brain organization that exhibit some stability quite early. Davidson *et al.* (1999) have also speculated that the frontal neocortical influence over affective tone operates via neocortical–amygdaloid circuits that are functionally asymmetrical. The left side of the frontal neocortex exerts a dampening influence on the amygdaloid nuclei, which respond to aversive stimuli and serve to sustain the response pattern associated with negative emotion.

Personality is usually evaluated by asking individuals about how they typically behave. In the case of young individuals, adults who frequently interact with them are asked to describe their behavior. The study of personality has a long history. Some of the major issues that have characterized this work include the following: (1) how best to determine the stable characteristics of a particular individual and (2) how many stable characteristics give sufficient detail about the individual to permit reasonable prediction of behavior and other responses across situations.

With respect to the number of abstract characteristics that permit an adequate representation of individual behavioral differences, the consensus that has been achieved argues for five major dimensions. These dimensions are known as the "big five" and capture the following aspects of individuality: (1) propensity to experience negative affect (neuroticism), (2) propensity to experience positive affect and be outgoing (extraversion), (3) interest in novel experiences (openness to experience), (4) agreeableness (ability to get along), and (5) conscientiousness (trust worthiness). These dimensions have been abstracted from extensive questionnaires concerned with a wide range of behavior in a variety of settings. In essence, even though individuals respond to hundreds of questions, their responses are sufficiently intercorrelated so that they can be adequately captured within the aforementioned five dimensions. It should be noted that what these questionnaires measure is the result of self-observations. One assumes they have some correspondence to neural dynamic properties but obviously cannot be identical with them. In other words, it is conceivable that two individuals who describe themselves identically have distinct patterns of dynamic brain organization. However, because the correspondence is assumed to be nonrandom, by knowing someone's personality profile, one achieves some constrain or prediction in the range of possible patterns of dynamic brain organization.

The most important idea to keep in mind in this context is that personality assessments give a bit of a handle on neural dynamic properties, and as such, they would be expected to have some degree of association with aspects of immune responsiveness. The literature in this regard is limited, but some interesting observations have been made. For instance, hostility is associated with enhanced NK cell activity, whereas both extraversion and dispositional optimism are associated with lower NK cell activity. One extensive study did not find a relationship between negative affectivity and NK cell activity, although right prefrontal activation, a correlate of negative affectivity, has been found to be associated with diminished NK cell activity. A number of other isolated findings have been reported, which suggest that the composition of peripheral blood leukocytes may be influenced by aspects of personality, particularly by the individual's affective profile. This, in turn, could affect functional measures of immune function, such as NK cell activity or lymphocyte proliferation.

X. CONCLUDING COMMENTS

Neural modulation of immune function is supported by a wide range of observations. First, there is a neural component to the endocrine effects on immune responsiveness. There are peripheral nerve endings in all the tissues where immune responses take place, as well as in the primary lymphoid tissues, where immune cells develop. The necessary molecular messengers, substances released by neural elements, have been found to operate by binding to receptors on leukocytes. Neuroanatomical evidence supports the notion that peripheral neuroimmune activity can be transmitted centrally and serves to organize higher order responses that appear in the form of behavior. In a similar fashion, the activity of higher order neural networks can influence microenvironments where immune cells are responding to antigen. Brain lesions, as high as the limbic system and neocortex, have detectable effects on immune responsiveness. Further evidence arises from the conditioning studies, which demonstrate that immune activity can be associated with neutral stimuli via the mediation of the CNS. Finally, the mounting evidence that personality, a manifestation of dynamic aspects of brain organization, has immune function correlates underscores the neuroimmune connection and supports early ideas that temperament or personality plays a role in disease.

XI. SOURCES

Ader, R., Felten, D.L., & Cohen, N. (eds) (2001), *Psychoneuroimmunology*, 3rd ed, San Diego, Academic Press.

Black, P.H. (2000), "Stress and the Inflammatory Response: A Review of Neurogenic Inflammation," *Brain, Behavior, and Immunity*, 16, pp. 622–653.

Cohen, F., Kearney, K.A., Zegans, L.S., et al. (1999), "Differential Immune System Changes with Acute and Persistent Stress for Optimists vs. Pessimists," *Brain, Behavior, and Immunity*, 13, pp. 155–174.

Cohen, J.J. (1999), "Individual Variability and Immunity," *Brain, Behavior, and Immunity*, 13, pp. 76–79.

Davidson, R.J., Coe, C.C., Dolski, I., & Donzella, B. (1999), "Individual Differences in Prefrontal Activation Asymmetry Predict Natural Killer Cell Activity at Rest and in Response to Challenge," *Brain, Behavior, and Immunity*, 13, pp. 93–108.

Felten, D.L., Cohen, N., Ader, R., et al. (1991), "Central neural circuits involved in neural-immune interactions," in Ader, R., Felten, D.L., & Cohen, N. (eds) *Psychoneuroimmunology*, 2nd ed, San Diego, Academic Press, pp. 3–25.

Gazzaniga, M.S. (ed) (2000), "*The New Cognitive Neurosciences*, 2nd ed, Cambridge, Mass, MIT Press.

Gershon, M.D. (1998), "*The Second Brain*, New York, Harper Collins.

Geschwind, N. & Behan, P.O. (1984), "Laterality, hormones and immunity," in Geschwind, N. & Galaburda, A.M. (eds) *Cerebral Dominance: The Biological Foundation*, Cambridge, Mass, Harvard Press.

Geschwind, N. & Galaburda, A.M. (1985), "Cerebral Lateralization: Biological Mechanisms, Associations, and Pathology: III. A Hypothesis and a Program for Research," *Archives of Neurology*, 43, pp. 634–654.

Hogan, R., Johnson, J., & Briggs, S. (eds) (1997), *Handbook of Personality Psychology*, San Diego, Academic Press.

Jänig, W. (1996), "Neurobiology of Visceral Afferent Neurons: Neuroanatomy, Functions, Organ Regulations and Sensations," *Biological Psychology*, 42, pp. 29–51.

Jurkowski, M., Trojniar, W., Barsman, A., et al. (2001), "Peripheral Blood Natural Killer Cell Cytotoxicity after Damage to the Limbic System in the Rat," *Brain, Behavior, and Immunity*, 15, pp. 93–113.

Kemeny, M.E. & Lauderslager, M.L. (1999), "Beyond Stress: The Role of Individual Difference Factors in Psychoneuroimmunology," *Brain, Behavior, and Immunity*, 13, pp. 73–75.

Maier, S.F. (2003), "Bi-directional Immune–Brain Communication: Implications for Understanding Stress, Pain, and Cognition," *Brain, Behavior, and Immunity*, 17, pp. 69–85.

Miller, G.E., Cohen, S., Rabin, B.S., et al. (1999), "Personality and Tonic Cardiovascular, Neuroendocrine, and Immune Parameters," *Brain, Behavior, and Immunity*, 13, pp. 109–123.

Sanders, V.M. & Straub, R.H. (2002), "Norepinephrine, the β-Adrenergic Receptor, and Immunity," *Brain, Behavior, and Immunity*, 16, pp. 290–332.

Vinken, P.J. & Bruyn, G.W. (eds) (1999), "The autonomic nervous system (part 1): Normal functions," in Apperzeller, O. (ed) *Handbook of Clinical Neurology*, vol 74, Amsterdam, Elsevier.

Yamada, T. (ed) (1999), *Textbook of Gastroenterology*, vol 1, Philadelphia, Lippincott Williams & Wilkins.

Zigmond, M.J., Bloom, F.E., Landis, S.C., et al. (eds) (1999), *Fundamental Neuroscience*, San Diego, Academic Press.

CHAPTER 6

Stress, Contextual Change, and Disease

I. Introduction   103
II. Selye's Concept of Stress   103
III. Ranking Life Events as Stressful   106
IV. Stress as Contextual Change   108
V. Social Context   109
VI. Other Life Forms   109
VII. Nonliving Environment   110
VIII. Individual as Context   111
IX. Disease as Contextual Change   113
X. Concluding Comments   114
XI. Sources   115

I. INTRODUCTION

This chapter focuses on the concept of stress in preparation for later chapters that consider evidence that stress has an impact on health. The notion of stress grew out of observations indicating that there were specific endocrine and immune system changes that were initiated by a wide range of noxious events including exposure to pathogens. Thus, by definition stress has endocrine and immune effects. Moreover, the range of events that can be subsumed under the rubric of stress is quite vast. This has been evident from the outset but is often not taken into account in popular discussions of the topic. In the following pages, the concept of stress is viewed as fundamentally denoting contextual change that demands a response. Context is considered a dynamic entity defined by the level and the focus of analysis, in other words, by where one chooses to set the boundary around a system.

II. SELYE'S CONCEPT OF STRESS

The term *stress* was first employed by Hans Selye in 1936 to denote the common response of organisms to a variety of noxious events, which included physical trauma (e.g., surgery and burns), chemical exposure (e.g., drugs), radiation, infections, and other challenging (e.g., intense exercise) or life-endangering (e.g., anoxia, low temperature) conditions.

Selye regarded the stress response as relatively nonspecific (i.e., independent of the particular type of noxious stimulation). He documented a number of physiological and tissue changes that occur with stress. He observed changes in blood electrolytes, glucose and lipid levels, blood pressure, and blood clotting time. The changes could be biphasic. For instance, initially hypothermia may be evident in response to stress, which then could give way to fever. The white blood cell count in the circulation may decrease initially and then progressively increase well above the prestress baseline. Gastrointestinal tract ulcers appeared and there were changes in glandular tissue, specifically hyperplasia of the adrenals and involution of the thymus. Thus, from the outset, stress has been associated with dramatic endocrine and immune changes.

Selye noted three phases in the stress response. There was an initial *alarm reaction*, a mobilization of the body's defenses. This phase had two components: an initial *shock-reaction* phase, from the offensive stimulation, soon followed by a systemic *counter–shock-reaction* phase, which redistributed organismic resources to counteract, neutralize, or overcome the challenge at hand. If the challenge persisted, the organism then entered a *resistance phase*, during which it appeared to handle the offending condition well and remained stable. However, if the noxious stimulation did not abate, the organism moved on to the *exhaustion phase*, wherein attempts to hold the challenge in check ceased and could result in the organism's demise. Selye referred to this dynamic response as the *general adaptation syndrome*.

Selye demonstrated that the effects of two distinct forms of noxious stimulation (i.e., two stressors), such as a cold environment and an infectious agent, depended on when the onset of each event occurred vis-à-vis the other. Essentially, if the second stimulus was presented during the shock reaction to the first, it augmented the organism's initial shock response. During the counter–shock-reaction phase to the first stimulus, the second stimulus had an attenuated impact. However, during the resistance phase to the first stimulus, the organism was much less able to handle the second stimulus and this persisted into the exhaustion phase, at which point, the organism began to lose its ability to manage the first stimulus. These observations led him to the concept of "adaptation energy." Such energy was a limited resource and would eventually become depleted when the onslaught of noxious or challenging stimulation was unrelenting.

Selye viewed *stress* as an all-encompassing term. By considering a wide variety of specific phenomena as threats or challenges to an organism, one can better understand how such events can overwhelm the organism. Stress is not solely a function of the psychosocial challenges confronting an individual. The impact of stressors depends on their number, their magnitude, and their temporal juxtaposition. Stress is superimposed on an original potential for adaptiveness. Individual adaptiveness is ultimately reflective of the history of the organism all the way back to genetic endowment, itself a reflection of each species' history.

Selye's notion of stress has been criticized for being overly abstract and overinclusive. This is understandable but it does not detract from the observation that a wide range of circumstances triggered by contextual changes provoke a mobilization of organismic resources. The issue of whether the term *stress* should be preserved for some level or degree of contextual change, rather than any change, however miniscule, has been difficult to resolve. The problem with attempting such a refinement is that it cannot be accomplished without regard to the individual. According to Berczi (1997), "The variation among individuals in their response often frustrated Selye in his attempt to find the perfect definition of stress, which, even today, we find elusive." This is not surprising given the complex and dynamic nature of contextual change. It is probably best to view any change as initiating stress and consider the ultimate impact of such change, with respect to whether it will have a beneficial or a detrimental effect, as only quantifiable in terms of conditional probabilities. The organism's ability to manage contextual changes is finite, and if enough challenges occur in a temporally proximal manner, then the organism's well-being can be compromised. As noted previously, a crucial aspect of the response to contextual change of any type is the condition of the individual at the moment of change. Consequently, there is marked variability in individual response. Some of this individual variation reflects not only relatively stable characteristics (e.g., genetic endowment, personality, and coping ability) but also situational differences such as whether exposure to a pathogen occurs during a period of examination, a time of vacation, or a period requiring only the execution of well-established daily routines.

Selye's work was primarily conducted with laboratory animals. However, his research has been very influential and served to popularize the role of stress in human disease. He wrote extensively about what he viewed as diseases of adaptation. He regarded these disorders as resulting from the damaging effects of adaptation responses that were excessive in magnitude, direction, and/or duration. He recognized, however, that in humans, the stress response gains a new order of complexity because of the mental prowess of humans, which through culture and its social institutions can dramatically expand the range of circumstances that qualify as threats or challenges.

The integrative aspect of Selye's work is emphasized here. He brought together under the rubric of stress diverse phenomena such as (1) toxicity (e.g., exposure to chemicals or radiation), (2) infection (e.g., colonization by microorganisms), (3) physical trauma (e.g., injury to tissue), and (4) adverse conditions (e.g., cold temperature, lack of oxygen, forced activity, and loss of sleep). He and others further demonstrated the impact of psychological manipulations (e.g., conditioned emotional responses and demanding cognitive performance) and social conditions (e.g., crowding, isolation, separation, and in the case of humans loss of a loved one).

The focus here is on stress as fundamentally the result of contextual change. Consequently, the discussion is organized in terms of four major dimensions of context: (1) the social setting (i.e., other interacting humans), (2) the other life forms in the environment (e.g., animals, insects, plants, and microbes), (3) the nonliving aspects of the setting (e.g., climate, altitude, radiation, air, and water), and (4) the characteristics of the individual as a feature of the context. With respect to the last dimension, it should be noted that when one examines any bodily function or activity within an individual, including mental activity, all else that transpires within the individual serves as part of the context for the function under scrutiny.

The position taken here is that any or all change initiates stress, which is not necessarily damaging. It is only when the organism's response proves ineffective in adapting to the change that the conditions for possible adverse outcomes may be set in motion.

III. RANKING LIFE EVENTS AS STRESSFUL

Holmes and Rahe (1967) have provided a framework for assigning numerical value to the degree of readjustment that people believe would be necessary to successfully adapt to a variety of life events. They quantified readjustment to both positive (e.g., marriage) and negative (e.g., jail term) life events. The procedure they employed began by arbitrarily assigning the value of 500 to marriage and then asking individuals to rank 42 other events, as requiring more or less readjustment. The list of life events is shown in Table 6.1 and includes death of spouse, divorce, jail term, retirement, and changes in aspects of day-to-day life. Subjects ranked

Table 6.1

The Holmes-Rahe Ranking of Life-Changing Events

100	Death of spouse
73	Divorce
65	Marital separation
63	Jail term
63	Death of close family member
53	Personal injury or illness
50	Marriage
47	Fired at work
45	Marital reconciliation

Table 6.1 (*Continued*)

45	Retirement
44	Change in health of family member
40	Pregnancy
39	Sexual difficulties
39	Gain of new family member
39	Business readjustment
38	Change in financial status
37	Death of close friend
36	Change to different line of work
35	Change in number of arguments with spouse
31	Mortgage or loan over $10,000
30	Foreclosure of mortgage or loan
29	Change in responsibilities at work
29	Son or daughter leaving home
29	Trouble with in-laws
28	Outstanding personal achievement
26	Wife begins or quits work
25	Change in living conditions
24	Change in personal habits
23	Trouble with boss
20	Change in work hours or conditions
20	Change in residence
19	Change in recreation
19	Change in number or type of church activities
18	Change in number or type of social activities
17	Mortgage or loan under $10,000
16	Change in sleeping habits
15	Change in number of family get-togethers
15	Change in eating habits
13	Vacation
12	Christmas
11	Minor violations of the law

only 6 of 42 events as requiring more readjustment (i.e., being more stressful) than marriage. All six are considered negative events.

Holmes and Rahe (1967) focused on change from the preexisting conditions and not on the psychological meaning, emotional response, or social desirability of the event. Essentially, they were concerned with contextual change independent of the individual's appraisal. The work of Holmes and Rahe, as well as that of others, has documented a correlation between onset or exacerbation of illness and the *amount* of total change in circumstances within a given period preceding the change in health status.

IV. STRESS AS CONTEXTUAL CHANGE

The work of Selye and that of Holmes and Rahe underscore that when one speaks of stress, stressors, or challenges, one is fundamentally referring to the fact that an individual's context has been altered. When viewed in this manner, stress (or what others [McEwen, 1998] have referred to as *allostatic load*) is unavoidable in life.

In the previous section, it was noted that the individual is an aspect of his or her own context. Therefore, even if one could imagine a relatively stable external environment, changes within the individual (e.g., those that occur with aging) would by definition alter context and generate stress. However, the more typical scenario involves contextual change over the short term while the individual remains relatively stable. Such contextual change occurs at many levels. There are relationship losses due to death or mobility. The arrival of new individuals not only alters the social milieu but may also create changes at the level of microscopic life forms, which may be residing within the newly arrived individuals. In other words, new arrivals may act as vectors for the introduction of novel life forms into a given context. There are also changes from year to year in the weather phenomena with consequences for the local ecology. Disruptions such as floods directly cause stress and can bring other life forms closer to humans and create threats or facilitate infection. Individual mobility of necessity can result in drastic alterations of context. There are changes in social networks, exposure to novel aspects of the physical environment and to a variety of life forms ranging from the microbial to plants, insects, and animals capable of posing hazards. Finally, the mental activity of individuals plays a role in how aspects of the external world ultimately affect the other bodily systems. For instance, thinking that one is constantly at risk of being killed because there are microbes or snakes in one's environment can create a level of threat that, though not realistic, nonetheless alters context in a way that may be detrimental to the individual. In addition, mental activity on a different time scale, by virtue of being the driving force behind cultural and technological developments, alters the world we inhabit and the context of human life and, thus, generates stress.

V. SOCIAL CONTEXT

Social context is the medium wherein human life is sustained, as we must mate to procreate. It is also essential for the protection of the young until they gain the skills and prowess to fend for themselves. Thus, shifts in social context can pose threat in a variety of ways. Such shifts can lead to loss of protection (e.g., when the young are separated from caregivers) or to loss of the opportunity to procreate (e.g., when a mating partner is lost). Changes in social context can occur as a result of the arrival of new individuals, who could pose a threat to the preexisting hierarchy and, thus, to life-sustaining resources.

Humans rank loss of close relationships (e.g., death of a spouse or divorce) as more challenging than other forms of contextual change. New relationships also introduce challenges, such as when one enters a marriage or when a child is born into a family. A particularly challenging circumstance for humans is the gradual loss that may occur as a result of chronic illness or disability affecting a family member. For instance, taking care of a severely retarded child (in which case, the loss has to do with potential never achieved) or a family member afflicted with a progressive dementia (in which case, the loss is of skill previously evident). Humans are unique in being subject to the influence of even more symbolic losses such as when a public figure or a celebrity dies. For the same reason, because humans have the ability to participate in imaginary relationships, some individuals can rely on fantasy to maintain a relatively satisfactory sense of well-being even in isolation.

VI. OTHER LIFE FORMS

Other life forms are a source of stress in the sense that we are all trying to survive. There is a complex balance in how life-sustaining resources are distributed. Humans are much less susceptible to macroscopic predators such as lions, although the experience of encountering such a creature can be exceedingly stressful, clearly because life is threatened. A much greater threat comes in the form of microscopic life in all its diversity, which if able to gain entry into the body can compromise the individual's viability. Whereas some pathogens are destructive, others are able to peacefully coexist. These differences are ultimately linked to the pathogen's own ability to remain viable and flourish irrespective of the survival of the host (i.e., infected individual).

Living vectors are organisms that transmit pathogens from one host to another. They are crucial in the balance that is struck between the various life forms. Among the best-understood vectors are a variety of insects (e.g., mosquitoes and tics) that are equipped to penetrate the skin in their own search for nutrition. They, thus, become convenient carriers for microscopic life forms that cause human diseases such as yellow fever, malaria, and Lyme disease. In addition, transmission of pathogens can be facilitated by cultural developments such as the aggregation of individuals in cities without adequate sanitation or even the concentration of the sick in hospitals.

The balance that is struck between the various life forms benefits from stability. For instance, people learn to avoid the dangers that might be posed by plant toxins (which serve to protect plants from predators, including humans). More generally, inhabitants evolve defenses against unavoidable adversaries and microorganisms tend to become less virulent. Contextual change becomes destabilizing, disturbs preexisting compromises, introduces new challenges, and is, therefore, stressful.

Human mobility has been regarded as a source of stress. However, it is not simply a matter of leaving behind familiar surroundings that are part of the nonliving world (e.g., houses and weather) or social context (e.g., friends and family). It is also the case that one may be entering an altered world of competing life forms, and at least temporarily one may be at a relative disadvantage. For instance, microorganisms not previously encountered may abound. Similarly, the arrival of other individuals, animals, or plants from distant places can introduce microscopic organisms or particles that could infect unprepared local inhabitants. This has often been cited as the basis for the epidemics that devastated the indigenous people of the Americas when European explorers began their colonization. In modern times the spread of organisms such as those causing influenza and acquired immunodeficiency syndrome (AIDS) has been facilitated by our enhanced mobility. In essence, our ability to maintain health is context dependent. Context has multiple layers; it is contextual change at any level that initiates stress.

VII. NONLIVING ENVIRONMENT

The nonliving environment encompasses features such as solar radiation, weather, altitude, chemical composition of the earth, and availability of water. These features of the environment serve as a backdrop and create the conditions to which life forms must adapt and within which they must coexist in ways that promote their viability.

Epidemiological investigation has documented that the probability of being afflicted with specific disorders is not uniform throughout the inhabited regions of the world. For instance, susceptibility to autoimmune disorders such as multiple sclerosis appears to be higher in populations living in less temperate zones. Of course, the possibility exists that such geographic variability is reflective of genetic inhomogeneity or diversity in the life forms that make up regional ecologies. However, the possibility that the physical environment plays a role should not be overlooked.

The nonliving environment has become much more complex because of industry and technology. Exposure to synthetic chemicals and various forms of radiation (e.g., from high-voltage power lines and now even cell phones) has received attention because of their potential to harm health. Such aspects of the nonliving environment have been cited often in connection with cancer. Exposure to synthetic chemicals originating in work environments or other environments in which toxic chemicals may be encountered (e.g., in battlefields) may play a role in conditions characterized by fatigue, poor ability to concentrate, and malaise.

VIII. INDIVIDUAL AS CONTEXT

The individual is an aspect of his or her own context. In other words, when one focuses on any particular bodily function or system, all else that transpires within the organism, including the psychological, provides the context for the function or system that is being examined.

The preceding chapters have described how the immune system mounts responses to potentially harmful microorganisms or to cells that have undergone malignant transformation. Such immune responses occur within the context of other bodily systems, which can modulate immune activity via a variety of molecular signals (i.e., hormones, neuropeptides, neurotransmitters, or cytokines). In turn, the activity of the immune system, through the action of cytokines, antibodies, and other mediators of inflammation, alters the context within which glandular, neural, and other tissues must operate. At the core of these intraorganismic transactions is the genome, which can be viewed as yet another facet of context. It essentially sets the stage for all other interactions, because it is only by modulation of genomic expression that the organism's structure and function can be maintained within the space that defines health, despite the multitude of challenges that constantly confront a life. Mutations alter the genetic context, often leading to detrimental effects but occasionally imparting benefits, which manifest as adaptive fitness and are the basis of evolution. Moreover, viruses are capable of altering the genetic context of cells within a given individual. Once inside a cell, viruses insert their own DNA or in the case of retroviruses manufacture DNA that then becomes integrated into the host's genome.

At any level of analysis, as soon as one focuses on any feature of the organism (e.g., specific gene, enzyme, cell type, organ, or system), the rest of the organism becomes an aspect of context. Moreover, the specific operation of the component or system under scrutiny cannot be understood without considering its context. It is also important to keep in mind that there are layers of context. In effect, one can view some aspects of the organism as embedded within layers of context that proceed from more proximal to distal aspects of the rest of the organism. Contextual change poses a challenge that varies depending on the contextual layers that have been penetrated. For instance, pathogens pose a risk if present in the immediate environment of an individual, but their potential to cause harm is greatly amplified if they enter the body. The same is generally true for toxins. Predators must be within striking distance of the individual without the benefit of a protective barrier (e.g., cage). Natural or human-made disasters must occur in the vicinity of an individual.

Pursuing the notion of the individual as an aspect of his or her own context brings one into a realm that as far as anyone can tell is uniquely human. This is the realm of consciousness, within which our knowledge (or absence thereof) serves in part to define our context. A vivid example of this phenomenon concerns ideas regarding human immunodeficiency virus (HIV) and AIDS. Some individuals believe that casual contact can transmit HIV and maintain a level of apprehension that is unwarranted. Others, who lack knowledge about the versatility of HIV, experience outrage at the fact that effective treatments or preventive interventions in the form of vaccines have not been found. Still other individuals appear convinced that HIV is not the cause of AIDS and are not prone to take adequate precautions (i.e., practice safe sex). Clearly, mental phenomena can alter context quite radically and in the latter case could increase the probability of infection. Even in the first two cases, the apprehensiveness or frustration brought on by ideas may lead to neuroendocrine changes that, in turn, could have adverse effects on the individual's health. Our conscious ideas, along with unconscious mental phenomena, serve to form our context.

Mental processes can also serve to counteract other contextual threats by a variety of evasive maneuvers that develop from an understanding of what is being confronted. In other words, a group can learn to avoid a particular food that induces illness, or one can develop a technology to defend oneself against predators of any sort. However, partial understandings can lead to unforeseen and detrimental consequences such as enhancing the lethality of the predator. For instance, the tendency to ignore the evolutionary perspective in efforts to eradicate pathogens has created strains that are more resilient. In effect, the mental prowess that led to the discovery of antibiotics altered the context wherein bacteria needed to survive and has led to bacterial adaptation in the form of antibiotic-resistant strains.

The preceding discussion serves to highlight that our mental life, what transpires within the human brain, is again a crucial aspect of the context of our existence and an important determinant of the challenges we must confront. Moreover, some important challenges to our well-being arise preponderantly within our mental life. One clear example of this phenomenon concerns the effects of undergoing examinations on bodily responses. Examinations constitute stressors only to the extent that we have aspirations to achieve in areas requiring competence that depends on formal education. Gearing up for an examination that could derail one's journey to a valued education-dependent goal poses a significant threat, which (as is described in the next chapter) has a detectable impact on the individual's ability to mount immunological responses that sustain health. Many laboratory situations used to induce stress in humans are effective only because of our mental activity. For instance, having to perform cognitively, as in the case of engaging in mental arithmetic, having to speak in front of others, or having to make rapid decisions proves challenging only because we care about our abilities or at least about how others will judge our competence. The role of mental activity as a determinant of context is obviously not without limits. It does not directly affect those aspects of the world that have an impact without having to be mediated by classical sensory systems or that cannot be avoided by behaviors that cause changes in one's environment.

Aspects of brain structure/activity, which fundamentally reflect genetic factors and the organism's history of experience or exposure, are reflected in mental processes. Essentially, previous contexts have shaped the brain and, thereby, have a role in determining the present context via the current mental processes. Brain activity contributes to an internal chemical milieu that modulates the immune system.

Previous chapters have outlined the pathways whereby the brain is able to interact with endocrine and immune systems. What is going on in the brain is part of what shapes the internal milieu. For instance, if one assesses any situation (e.g., public speaking) as a threat in some sense, one's body will respond as if the confrontation was with a predator. Through a sense of invulnerability, mental activity can also cause an individual to risk exposure to pathogens by engaging in unprotected sex or some other high-risk behavior. How mental activity can be harnessed to shape context in a way that sustains health is the topic of Chapter 11.

Mental activity becomes a factor in whether and when one seeks contact with others, and this may be gender dependent to some extent. Individuals may isolate themselves and the lack of social stimulation may trigger changes in neuroendocrine activity, which may jeopardize health. Specifically, although isolation can lower the risk of some types of infection, it may render the individual more vulnerable to reactivation of latent infections, malignant transformations, or other problems such as cardiovascular dysfunction. Mental activity is further implicated in health by how it facilitates or hinders seeking medical assistance when functional difficulties arise. However, some individuals are driven to seek medical consultation to their detriment. It is quite evident that what transpires in the province of the mind is an important determinant of context and, therefore, health.

IX. DISEASE AS CONTEXTUAL CHANGE

A disease is a form of contextual change and, thus, fundamentally stress. Infectious diseases occur when pathogens invade and grow in an organism. Their presence within the body alters the context of bodily tissues. Allergic conditions arise in a similar way. Typically, some aspect of the environment (e.g., pollen) gains entry into the body and in susceptible individuals induces tissue responses that can be distressing and potentially damaging to the organism. Disorders such as cancer can be initiated by infectious agents that alter the genetic context of cells and may promote malignant transformations. Autoimmune disorders can also be triggered by infections, which through molecular mimicry mechanisms may cause immune attack of healthy self-tissues. Other aspects of context, such as toxic chemicals or even radiation, can, via effects on enzymes, receptors, or directly on the genome, cause cell changes that may unleash malignancy or autoimmunity. Such disorders also can arise without evidence of any specific pathogen or other type of insult. They can simply emerge from the complex interactions among sets of genes, which need not even be identical across individuals. Such complex scenarios are consistent with the failure to find relatively simple causes for disease. They underscore the need to adopt approaches that consider all aspects of contextual change as underlying stress and, therefore, being pertinent to health maintenance.

X. CONCLUDING COMMENTS

Selye's work established the concept of stress as an organismic response to a wide range of events that are noxious or potentially challenge the organism's integrity. More generally, the stress response occurs as a result of contextual change, which is simply unavoidable in life. Thus, stress is ever present.

Selye emphasized that the response to contextual change is complex and depends on the organism's ability to counteract the change, which is further complicated by the type of change, the amount of change, and the extent to which changes overlap in time. Moreover, the response to change requires resource expenditure, that is, use of finite "adaptational energy." Thus, the organism's ability to manage stress has limits.

What constitutes context depends on where one draws the line or what one chooses as the focus of interest. Aspects of the individual become context as one enters the body and progressively focuses on specific systems or structures all the way to the level of molecules. The individual's genetic endowment and his or her mental processes can be viewed as context. The individual shapes context constantly, at all levels, by how he or she thinks, feels, and behaves over time. Thus, stress is a highly individual response.

Obviously, context extends beyond the individual's contribution and can be conceptualized in terms of its social composition, its diversity with respect to other life forms, and its nonliving characteristics. These dimensions of context are dynamic and subject to the influence of forces outside of the individual. Thus, stress is not simply reducible to individual factors.

In essence, context is constantly changing, due to both individual outlook/ behavior and individual circumstances. The individual's responses can prove effective in handling the resulting challenges, so stress can promote resilience and adaptiveness. However, an individual's coping resources are limited and, therefore, can be overwhelmed with the consequence that the internal state that defines health would be jeopardized. Thus, stress can have both beneficial and adverse effects.

XI. SOURCES

Berczi, I. (1997), "The Stress Concept: An Historical Perspective Hans Selye's Contributions." In J.C. Buckingham, G.E. Gilles, and A.-M. Cowell (Eds.) *Stress, Stress Hormones, and the Immune System.* Chichester, England, John Wiley & Sons, pp. 1–6.

Diamond, J. (1997), *Guns, Germs, and Steel. The Fates of Human Societies.* New York, W. W. Norton.

Ewald, P.W. (1994), *Evolution of Infectious Disease*, New York, Oxford University Press.

Holmes, T.H. & Rahe, R.H. (1967), "The Social Readjustment Rating Scale," *Journal of Psychosomatic Research*, 2, pp. 213–218.

McEwen, B.S. (1998), "Stress, Adaptation and Disease: Allostasis and Allostatic Load," *Annals of the New York Academy of Sciences*, 840, pp. 33–44.

McMichael, T. (2001), *Human Frontiers, Environments and Disease: Past Patterns, Uncertain Futures*, Cambridge, United Kingdom, Cambridge University Press.

Nesse, R.M. & Williams, G.C. (1994), *Why We Get Sick: The New Science of Darwinian Medicine*, New York, Vintage Books.

Selye, H. (1936), "A Syndrome Produced by Diverse Nocuous Agents," *Nature*, 138, p. 32.

Selye, H. (1946), "The General Adaptation Syndrome and the Diseases of Adaptation," *Journal of Clinical Endocrinology*, 6, pp. 117–230.

Selye, H. (1956), *The Stress of Life*, New York, McGraw-Hill.

Starling, P. & Eyer, J. (1988), "Allostasis—A new paradigm to explain arousal pathology," in Fisher, S. & Reason, J. (eds) *Handbook of Life Stress, Cognition, and Health*, New York, John Wiley & Sons.

Taylor, S.E., Klein, L.C., Lewis, B.P., et al. (2000), "Biobehavioral Responses to Stress in Females: Tend-and-Befriend, not Fight-or-Flight," *Psychological Review*, 107, pp. 411–429.

CHAPTER 7

Psychosocial Stress:
Neuroendocrine and Immune Effects

I. **Introduction** 117
II. **Psychosocial Stress** 118
  A. Life Events 118
  B. Individual Attributes/Personality 119
  C. Laboratory Paradigms 119
III. **Effects on Endocrine Activity** 119
  A. Pituitary–Adrenal Axis 120
  B. Pituitary–Gonadal Axis 121
  C. Other Axes and Hormones 121
IV. **Effects on Autonomic and Peripheral Neural Activity** 122
  A. Classical Neurotransmitters 122
  B. Neuropeptides 122
V. **Effects on the Central Nervous System** 123
VI. **Effects on the Immune System** 123
  A. Nonspecific Immunity 124
  B. Humoral Immunity 125
  C. Cell-Mediated Immunity 126
  D. Cautions and Integration 128
VII. **Neuroendocrine–Immune Pathways** 129
VIII. **Concluding Comments** 129
IX. **Sources** 131

I. INTRODUCTION

The preceding chapters have delved into how the immune system mounts responses and how such responses appear to be modulated by the activity of hormones, peptides, and neurotransmitters released by endocrine and neural tissues. The notion of stress has also been examined and viewed as rooted in contextual change, the organismic response to such change, and the effectiveness of the response with respect to maintaining adaptiveness. Stress is produced by changes of any type and is, thus, a broad concept.

This chapter focuses on psychosocial stress and how it affects endocrine, neural, and immune activity, particularly in humans. The work on lower organisms is considered only in passing, because the goal is to highlight what has been observed in humans. This overview is not exhaustive; it is intended to highlight how efforts to promote human health and well-being may need to proceed. Moreover, the focus on human research should help make clear which areas need further investigation. This will facilitate later discussion of how such investigation can be performed without placing individuals at undue risk simply for the purpose of gaining knowledge.

II. PSYCHOSOCIAL STRESS

The social context of humans has received much attention with respect to how changes therein induce stress. In fact, there is often a tendency to equate stress with social conditions and overlook the fact that *stress* is a much more encompassing term.

The psychological characteristics of individuals are part of what determines how changes in social and nonsocial conditions affect the individual's psychological equilibrium. Therefore, whenever data permit, discussion of psychosocial stress is organized in terms of the impact of life events or laboratory manipulation on individuals with specific psychological attributes, or personality traits.

A. LIFE EVENTS

Life events occur as a result of the natural course of life. They encompass the occurrences listed on the Holmes and Rahe social readjustment scale and include catastrophic developments that disrupt the individual's environment, both naturally occurring (e.g., earthquakes) and human-made (e.g., war).

Life event studies are necessarily correlational, as researchers have no control over when and to whom the events occur. An exception to this general rule occurs in the case of events that can be planned, such as when individuals train in potentially hazardous activities (e.g., parachute jumping) or engage in challenging encounters such as sporting events, academic examinations, or public performances. However, individuals are not typically randomly assigned to such challenges. In fact, random assignment of representative samples to experimental manipulations is an ideal that is seldom achieved in human research.

B. Individual Attributes/Personality

As was noted in the previous chapter, the individual can be viewed as an aspect of context. Therefore, objective events, such as the death of a spouse or an examination, are not the sole determinants of the impact on bodily systems within the individual. Characteristics of the individual at various levels are also factors in what occurs.

Typically, studies assign psychological attributes or personalities to subjects based on how they respond to a series of standard questions. Less frequently, such characterization is based on inferences from the content of stories made up by the individual in response to standard scenarios. The assumption here is that the individual endows imagined characters with his or her own concerns, issues, aspirations, or fundamental outlooks. Occasionally, the individual may be characterized based on ratings or observations of behavior made by others. Unfortunately, studies do not routinely examine both individual characteristics and circumstances surrounding the individual, thereby missing the opportunity to observe the interaction between stable individual attributes and life events.

C. Laboratory Paradigms

Researchers contrive situations that are assumed to challenge subjects mentally (e.g., mental arithmetic or rapid decision making) or that cause distress as a result of either aversive stimulation (e.g., electric shock) or social discomfort (e.g., public speaking). By definition, laboratory situations are artificial and, thus, lack the significance of real-life challenges. Nonetheless, they retain some measure of challenge. Moreover, it is possible to randomly assign subjects to conditions, but still generalizations are possible only with respect to the population of those who volunteer. Volunteers are not necessarily representatives of all sectors of humanity. Thus, in the human arena, one is always left with the nagging concern that one has only a piece of the puzzle. This is clearly better than pure guesswork but still demands that one proceed cautiously, particularly with respect to groups that have not been adequately represented in the research.

III. EFFECTS ON ENDOCRINE ACTIVITY

The research has not been sufficiently extensive to examine in any detail the various endocrine axes that would be expected to modulate immune activity, as described in Chapter 4. Studies have addressed changes in only one or a few of the endocrine axes. Thus, in general, there is only a fragmentary picture of how psychosocial changes, which generate stress, alter endocrine function. Nonetheless, evidence already clearly suggests that simplistic notions of stress effects must be discarded.

A. Pituitary–Adrenal Axis

Activation of the pituitary–adrenal axis is a defining characteristic of stress, although individual differences in the extent of such activation exist. The research has shown that a variety of stress-inducing situations, both transient and chronic, are associated with increased release of adrenocorticotropic hormone (ACTH) along with β-endorphin and with elevated cortisol levels in the circulation. Such increases may at least temporarily amplify the effect of the pituitary–adrenal axis on other tissues. However, because studies have not examined whether there are coincident changes in cortisol-binding proteins or cortisol receptors, the actual impact of increased cortisol remains speculative. In fact, when cortisol levels are assessed along with measures of immune responsiveness, the correlation is often not significant. This can occur when the activity of a system is multivariately determined and not enough variables have been considered. The evidence has also shown that when a stress-producing situation is mastered, cortisol levels return to baseline. For instance, this has been specifically observed when soldiers undergo paratrooper training. Initially, cortisol levels increase dramatically, but with experience, the increase ceases to occur.

The research on primates nicely illustrates the impact of social context. Subjects who have a lower rank in the social group have higher levels of ACTH and cortisol. These studies also suggest that cortisol receptors may be downregulated, at least in some tissues, so that cortisol is not able to inhibit the activity of neural structures that promote its release. It should be noted that if receptors are downregulated across tissues, the elevated cortisol levels would not be functionally significant. Only tissues that retain cortisol receptors at baseline levels will show an altered response.

The influence of social context is complicated by the fact that under some circumstances, maintaining dominance is much more challenging than in other situations, and thus, the cortisol relationship to dominance can disappear or reverse. The experience of individual animals can be drastically different even within a relatively stable hierarchy. For instance, an individual of intermediate rank that is subjected to frequent challenges by lower ranking individuals will tend to have a higher cortisol level than the lowest ranking individual. Moreover, the "personality" of a dominant animal will make a difference as well. Those that are more readily aggressive upon the slightest hint of threat and who succeed with their aggression have the lowest cortisol level, even within an all-dominant group.

In addition, the extent to which animals interact physically with others, as occurs during grooming, seems to have a cortisol-lowering effect. Clearly, individual factors and contextual features moderate the effect of stress-inducing circumstances with respect to pituitary–adrenal activity. However, generally such stress-inducing situations cause an increase in adrenocortical activity and the extent of the increase tends to be more pronounced in male organisms.

B. PITUITARY–GONADAL AXIS

Stressors such as parachute jumping or undergoing examinations are most often associated with inhibition of gonadotropin-releasing hormone (Gn-RH) release. This effect may be mediated through activation of the hypothalamic–pituitary–adrenal (HPA) axis, which, in turn, inhibits Gn-RH release, possibly through the effects of opioid peptides. Inhibition of Gn-RH release leads to decreased luteinizing hormone (LH) and follicle-stimulating hormone (FSH) levels in the circulation and ultimately reduced estradiol in females and testosterone in males. Thus, stress appears to cause the decrease of gonadal steroids.

C. OTHER AXES AND HORMONES

The work with respect to the pituitary–thyroid axis is more limited. TSH levels are responsive to intense stressors. Frequently the response is a decrease in TSH levels, although in some situations, TSH release is increased. For instance, parachute jumping tends to increase TSH release at least in some individuals. Changes in TSH release are, in turn, reflected in the level of thyroid hormones in the circulation. Growth hormone (GH) release appears to be increased in humans during the acute phase of a stress-inducing situation. However, when situations produce chronic stress, GH release may be inhibited. This appears to be true in the case of persistent maltreatment and deprivation directed at young children, who have been found not to grow normally, a condition known as *psychosocial dwarfism*. Prolactin (PRL) release tends to be increased by acute stressors, but the extent of its increase appears dependent on the baseline PRL level and the concomitant level of cortisol. The higher both of these are, the lower the PRL response to acute stress. The posterior pituitary hormones, vasopressin and oxytocin, both tend to increase in the circulation at least initially in response to stress.

Overall, it appears that most pituitary hormones exhibit a biphasic pattern of release in response to stress. They initially increase in the circulation and then return to baseline or may even become suppressed. This is what is expected when release is under negative feedback by products released from target glands or other tissues. The pituitary hormones may serve to mobilize the organism's metabolic resources and, thereafter, gradually fade into the background.

IV. EFFECTS ON AUTONOMIC AND PERIPHERAL NEURAL ACTIVITY

Stress causes changes in the activity of the autonomic nervous system (ANS) and more generally the peripheral nervous system (PNS). The research demonstrating such changes has focused on classical neurotransmitters, particularly norepinephrine (NE) and epinephrine (E) (adrenaline). However, the evidence for colocalization of peptides in ANS and PNS terminals, at least indirectly, implicates peptides in the ultimate impact of stress on a variety of tissues.

A. CLASSICAL NEUROTRANSMITTERS

Stress activates the sympathetic branch of the ANS and promotes the release of NE and E from nerve terminals and the adrenal medulla, respectively. Acetylcholine (ACh) release from the sympathetic postganglionic neurons that innervate the sweat glands is also increased. ACh release from the parasympathetic branch of the ANS is decreased, because parasympathetic activity tends to be suppressed for the most part. The fact that classical neurotransmitters such as NE and E are colocalized with various peptides, including the enkephalins, raises the possibility that such peptides serve to modulate tissue responses to stress as well. For instance, there is evidence that pain stimulation produced by electric shocks can induce analgesia. This effect depends on the integrity of the adrenal medulla, and it can be prevented by drugs that block the action of the enkephalins.

B. NEUROPEPTIDES

The research focusing on peptides has been less extensive. As noted earlier, stress is expected to cause changes in the levels of peptides that are co-released with the classical neurotransmitters. For instance, the available evidence suggests that individuals who report high anxiety when confronted with situations that are potentially dangerous (e.g., parachute jumping) or uncomfortable (e.g., colonoscopy) have higher levels of substance P in the circulation, which can arise from a wide range of sources, but its colocalization with classical neurotransmitters in the PNS makes it likely that its elevation reflects, at least in part, increased neural activity.

V. EFFECTS ON THE CENTRAL NERVOUS SYSTEM

Animal studies have documented that stress-inducing conditions cause wide-spread activation of structures within the central nervous system that are known to play a role in emotion, ANS responses, and endocrine regulation.

Endocrine activity, particularly that of the adrenal gland, modifies neuro-chemical systems in the brain. In fact, glucocorticoids seem initially to operate as anxiolytic or antidepressive molecules, most likely through their action on the monoaminergic and GABAergic neuronal systems and through their inhibition of neurons that release corticotropic-releasing hormone. Glucocorticoids released in response to stressful events initially enhance GABAergic transmission but over time cause downregulation of γ-aminobutyric acid (GABA) receptors. Glucocorticoids counterbalance the NE-stimulated production of cyclic adenosine mono-phosphate, thus attenuating the effect of increased NE release, particularly in the neocortex.

Stress increases 5-hydroxytryptamine (5-HT; serotonin) utilization and glucocorticoids help maintain 5-HT synthesis at a high enough rate to sustain a constant level of 5-HT. Some 5-HT receptors appear to be upregulated by stress, but high levels of glucocorticoids decrease binding to 5-HT receptors, particularly in the hippocampus.

The hippocampus appears especially sensitive to the effects of glucocorticoids. Neurons in the hippocampus appear to undergo apoptosis (cell death) when sustained activity occurs in the presence of high glucocorticoid levels. Damage to the hippocampus, in turn, promotes sustained HPA axis activation, creating a potentially self-reinforcing process. Something like this has been documented in patients with Alzheimer's disease who show evidence of hippocampal shrinkage. Thus, glucocorticoid activity, which is thought to have a counter-regulatory effect in response to stress (i.e., downregulate the alarm reaction) can at times have tissue-damaging effects and, thereby, alter the structure and function of the brain.

VI. EFFECTS ON THE IMMUNE SYSTEM

Psychosocial stress is predominantly transmitted to the rest of the organism via its effect on the brain. The research, as outlined earlier in this chapter, is sketchy with respect to the multitude of endocrine and neural pathways capable of mediating the impact of psychosocial stress on immune function. In particular, many of the suspected molecular mediators have not been sufficiently examined to permit a satisfactory characterization of their role *in vivo*. Nonetheless, a substantial body of evidence has accumulated concerning psychosocial stress and measures of immune function. The observed effects underscore the role of stress in immune-mediated health maintenance, even though the pathways transmitting such an influence remain insufficiently understood.

A. NONSPECIFIC IMMUNITY

Research on nonspecific immunity in humans is overwhelmingly focused on natural killer (NK) cells. The observations concerning other aspects of nonspecific immunity such as phagocytosis are limited. Chronic stress and serious depression have been consistently associated with increased neutrophil counts in the peripheral circulation. At least one study has reported decreased phagocytosis in response to laboratory stress and another has found reduced phagocytosis in children reporting depression symptoms and more life stress. Anxiety level in conjunction with how the individual copes with it has been positively correlated with monocyte and eosinophil counts. Moreover, individuals who amplify stress (high reactors) have increased neutrophil and monocyte counts. Overall, the white blood cell count is associated positively with level of negative affect.

Research has shown that NK cells in the peripheral circulation are increased after acute laboratory stress (e.g., speech task, mental arithmetic, puzzle solving, or electric shock), particularly in individuals with low levels of prior stress. NK cells are also increased after physical exercise and after living through a natural disaster such as a hurricane. However, in a study that took individual psychological differences (i.e., high vs low tendency to worry) into consideration, researchers found that after an earthquake (a natural disaster that is much less predictable than a hurricane), the high worry-prone individuals had a lower NK cell number. NK cell numbers have also been found to be lower in groups with high chronic stress levels and daily hassles, as well as in those who were caring for a permanently disabled relative or experiencing significant job-related stress. In addition to affecting NK cell numbers, stress alters the cytotoxicity, or cell-killing proficiency, of NK cells. NK cell activity is diminished during bereavement, during important examinations, in response to sleep deprivation, or after an angry exchange with one's spouse. Increased NK cell activity has been reported after some laboratory challenges. NK cell activity is increased in individuals living in war situations. Such increases are also evident when individuals anticipate pain. Individuals who tend to perceive stress as low, even when it is objectively high, show higher NK cell activity. In contrast, individuals who described themselves as depressed or lonely have reduced NK cell activity, even though being socially introverted is associated with increased NK cell activity. Because the sample populations used in the aforementioned studies included different types of individuals, the extent to which the preceding statements can be generalized is not entirely clear. For instance, the NK cell activity association with loneliness was found in a sample of psychiatric patients, whereas the relationship to social introversion was found in a sample of airline crew members.

Overall, the evidence appears to suggest that NK cells increase in the circulation in response to acute stress, but when the stress becomes chronic, they tend to become progressively less numerous. The activity of NK cells may be increased acutely for some individuals (e.g., low stress reactive) or in some situations (e.g., anticipation of pain), but again, there is evidence of diminished activity under conditions of more prolonged or unpredicted stress. Other components of nonspecific immunity such as phagocytosis also appear to decline with prolonged stress even though phagocyte numbers in the peripheral circulation tend to be increased. These inverse trends between increased cell number in the circulation and diminished activity could reflect mobilization of nonspecific immunity in preparation for possible injury while maintaining the prestress level of functional activity, at least for most individuals. In other words, the increased number of cells compensates for the diminished activity, thereby sustaining functional competence. However, the evidence further suggests that with prolonged stress, both NK cell numbers and activity become reduced, leading to impaired immune competence and susceptibility to otherwise improbable illness.

B. HUMORAL IMMUNITY

Humoral immunity is an aspect of specific immune responses directed at particular antigens. It takes the form of unique antibodies produced by B lymphocytes that have been specifically selected to neutralize the antigen at hand. The production of specific antibodies by activated B lymphocytes (plasma cells) occurs within lymphoid tissues; thus, the observation that B cells are reduced in the peripheral circulation in response to stressors, such as during space flight, does not allow straightforward inferences regarding whether humoral immunity has been suppressed or enhanced (i.e., most cells could have migrated into lymph nodes and are, therefore, poised to produce antibodies).

A number of studies have focused on secretory immunoglobulin A (sIgA), which, though specific with respect to antigen, participates in the blockade of pathogen entry through mucous membranes. In essence, it enhances nonspecific responses designed to block pathogen penetration. Secretory IgA in saliva is lower during a period of stress (e.g., an examination) compared to a baseline period. Individual characteristics determined based on psychological tests, such as "inhibited need for power" or "high external locus of control," are associated with a low sIgA level in saliva. In contrast, serum IgA level is increased during examinations, as well as in individuals who suppress anger. Thus, even though IgA is being produced, it may not be adequately transported to the mucosal surfaces, where it is needed, and therefore not found in secretions to the same degree as during stress-free periods or in less emotionally tense individuals.

Two other lines of research are pertinent to the influence of stress on humoral immunity. One set of studies shows that under a variety of chronic stressors (e.g., divorce or caring for a demented family member), the specific antibody titers to some latent viruses (e.g., Epstein–Barr virus [EBV], herpes simplex virus [HSV], or cytomegalovirus [CMV]) are elevated. At first glance, this seems to be evidence of enhanced activity, but researchers have concluded that it is a consequence of impaired cell-mediated immunity and, thus, an indication that the immune system as a whole is not working effectively. What is occurring is that as a result of impaired cell-mediated immunity, increased viral replication stimulates the production of specific antibody, a much less effective response to latent viruses.

The most clear-cut evidence of diminished humoral immunity comes from studies of response to vaccination with specific antigens, although again the interdependence of humoral (antibody production) and cell-mediated (T-cell help) responses may be a factor. Typically, vaccination with antigens such as hepatitis B produces lower titers in individuals who describe themselves as "under high stress" or report low levels of social support. This sort of observation has been more likely when the subjects are older and confronted with chronic stress. Younger subjects do not always show a relationship between perceived stress and response to vaccination.

In essence, interpretation of the significance of antibody levels must be done carefully. High levels of IgA in the blood may reflect a functional deficit, at least temporarily, because IgA performs its primary function when released in mucosal secretions. Similarly, high antibody titers to latent viruses may reflect increased viral replication due to impaired cell-mediated immunity. Low response to immunization could reflect some deficit in B-cell function but could also be a function of impaired activity in antigen processing or the T-cell help that is often required for antibody production.

C. CELL-MEDIATED IMMUNITY

The focus here is primarily on measures of T-cell function or integrated immune responses such as the delayed-type hypersensitivity (DTH) reaction, which depends on antigen presentation to helper T cells to mount both nonspecific (inflammation) and specific immune responses.

Studies of cell-mediated immunity have often examined the number of helper T cells (CD4$^+$) and cytotoxic T cells (CD8$^+$) in the peripheral circulation, as well as the ratio of CD4$^+$ to CD8$^+$ lymphocytes. Frequently, mitogen-induced proliferation has been investigated. Other measures that have been examined include DTH responses to previously encountered antigens and the predominance of cytokine profiles that would be expected to favor cell-mediated (Th1) as opposed to humoral (Th2) immune responses.

A variety of stressors such as a high number of daily hassles and natural disasters are associated with lower T-cell numbers in the peripheral circulation. However, this is not always the case. There are an equal number of instances in which events such as experiencing an earthquake, starting school, or experiencing marital conflict are associated with increased T-cell numbers, particularly helper T cells. Changes in T-cell number will often alter the $CD4^+/CD8^+$ ratio (i.e., helper-to-cytotoxic predominance). Again, the same event can shift the ratio up or down and possibly act to increase variability or diversity. Some types of stress (e.g., space flight or entering college) that can be regarded as positive challenges tend to be associated with an increased $CD4^+/CD8^+$ ratio, most often due to increased $CD4^+$ counts. In contrast, conditions such as caring for chronically ill relatives, job stress, a variety of laboratory challenges (e.g., mild shock, mental activity, or public speaking), and real-life cognitive challenges (e.g., undergoing examinations) are all consistently associated with a decrease in the $CD4^+/CD8^+$ ratio, most often the result of an increased $CD8^+$ count in the peripheral circulation.

T-cell proliferation, which is particularly dependent on the $CD4^+$ subpopulation, is most often decreased by a wide variety of stressors. Diminished T-cell proliferation has been observed in the unemployed, the bereaved, and those caring for sick or vulnerable individuals, particularly when the caregiver tends to be anxious. The range of conditions shown to be related to decreased T-cell proliferation is relatively large and includes space flight, threat of missile attack, entering school, undergoing examinations, marital discord, sleep deprivation, awaiting surgery, general life demands, and various laboratory conditions that require focused cognitive activity or social interaction with strangers (e.g., public speaking). There have been a few instances in which some of the aforementioned situations produced no change (e.g., space flight) or even an increase in proliferation (e.g., surviving an earthquake). High social support has also been associated with increased T-cell proliferation.

Finally, the DTH response appears enhanced in individuals who exhibit a high degree of social engagement. Stressful conditions such as awaiting surgery or undergoing examinations tend to diminish the DTH response. In addition, it has been observed that stress can shift the profile of cytokine release away from that necessary to promote cell-mediated immunity (Th1 responses) to that which facilitates humoral immunity (Th2 responses), specifically stress decreases interferon-α (IFN-α) and interleukin-2 (IL-2) while increasing IL-4 and IL-10.

D. CAUTIONS AND INTEGRATION

It is important to underscore the need for caution in interpreting the significance of the observations that have been summarized in this chapter, because study of immune cells and their activity in humans is restricted to cells found in the peripheral circulation. It is not reasonable to assume that what they reveal would apply to other tissue compartments such as lymph nodes, the spleen, or gut-associated lymphoid tissue. Nonetheless, the picture obtained from the peripheral circulation, though not necessarily descriptive of the system in all compartments, may be sufficiently informative to allow prediction of who will be at increased risk of illness and which type of illness may be more likely to manifest.

The preponderance of the findings appears to suggest that when the individual is confronted with stressful situations, immune cells in the peripheral circulation tend to increase, especially NK cells and CD8$^+$ cells. If the stress is sustained, there may be a decrease in all types of lymphocytes and an associated increase in phagocytes. Generally, the trend is for leukocytes to be functionally less active under stress. Thus, even though there are more NK cells, they appear to be less cytotoxic, making it possible that the changes (i.e., increased number of cells with decreased activity) simply maintain the prestress competence while mobilizing the leukocyte population.

The findings regarding humoral immunity seem closely intertwined with those regarding cellular immunity. Impairments in cellular immunity appear to be the hallmark of exposure to stress; this, in turn, can cause a failure to mount responses to novel antigen and reliance on humoral immunity to defend against previously encountered antigens, which may not be entirely effective and may in some instances create problems by attacking healthy tissues or innocuous substances. Another aspect of humoral immunity that appears adversely affected by stress, most likely independently of effects on cell-mediated immunity, is the secretion of IgA into mucosal fluids. This hampers the ability of the body to rid itself of some pathogens before they gain entry into the body. Cell-mediated immunity appears relatively dampened with respect to processing novel antigen but may remain reasonably poised to deal with previously encountered antigen. The latter occurs via the mobilization of memory T cells with the ability to be specifically cytotoxic, though in an attenuated manner because of the stress-induced downregulation of activity.

One point that cannot be overstated is that there are individual differences in the response to stress. At times, those who experience the least stress or are less reactive to stress respond to a stressor differently than other individuals exposed to the same stressor. This is consistent with the notion that stress interacts with the preexisting level of activity in any physiological system, ultimately leading to nonlinear responses. In a few instances, having lived through a significant stressor (e.g., space flight or earthquake) tended to be associated with enhanced immune responsiveness. Thus, individual factors, unique aspects of the situation at hand, and the state of the bodily system under scrutiny just before some event occurs will all shape the response.

VII. NEUROENDOCRINE–IMMUNE PATHWAYS

A variety of stress-inducing situations (via their effect on patterns of brain activity that remain poorly characterized) have consistently been found to increase the activity of HPA axis, diminish the release of gonadal hormones, particularly testosterone, and have biphasic effects (an initial increase followed by a return to baseline or a decrease) on hormones such as GH or PRL. Sympathetic nervous system activation also occurs in response to stress and is associated with increased release of catecholamines and colocalized peptides. At all levels, the influence is multidirectional: Contextual change alters brain activity, which regulates endocrine and autonomic activity, which modify brain activity and the function of other tissues including the immune system, all of which in their own right further modulate the brain activity underlying mental states and behavior, thereby having an impact on the context of the individual, which then begins the cycle all over again.

The available evidence supports the view that alterations in neuroendocrine activity have an impact on immune function. There appears to be an initial mobilization of leukocytes, along with a dampening of their activity, particularly regarding measures of cytotoxicity and proliferation. Antibody production does not seem to diminish; it may even increase, although the release of antibodies onto mucosal surfaces may be curtailed. Overall, the functional shift appears to be initially toward mobilizing the system while keeping it in check by dampening its reactivity and conserving resources. The impact is likely to be most pronounced with respect to mounting effective responses against novel antigens.

The response pattern described earlier should be considered the modal response and it should be anticipated that dispersion will occur as well and may even be increased. In other words, there will be subgroups that will respond different than the majority, at least with respect to one or more of the aforementioned measures. This should be the case because the changing nature of circumstances may make what has been previously adaptive for the population ineffective. Therefore, the population must retain diversity, even if it does not seem entirely adaptive at a given point or for all individuals, so its ability to cope with unforeseeable developments is not degraded.

VIII. CONCLUDING COMMENTS

Human studies are opportunistic, because one cannot employ random assignment to experimentally controlled conditions in humans, as can be done in the case of laboratory animals. However, it is the human responses to stress that are ultimately of interest with respect to safeguarding the health of individuals.

A major problem with human studies is that types of stressors and sample characteristics are frequently confounded. For instance, examinations typically affect young individuals who are intellectually capable. Space flight is reserved for unusually adventuresome and technically skilled individuals. Loss of a spouse is more likely in older groups. Certain types of disasters occur in geographically defined areas, where inhabitants recognize that they are at increased risk. Therefore, findings are not exactly generalizible. Seemingly, contradictory observations may reflect the fact that what the studies demonstrate are "cohort," "local," or "type-of-event" effects. For instance, hurricanes and earthquakes are both natural disasters but clearly differ in predictability and duration.

The evidence concerning endocrine modulation of immune function due to stress suggests that there is more to it than simply the action of the HPA axis. However, characterization of the role of other hormones is much more fragmentary, as is the examination of interactions among the major hormones. For instance, the effect of high cortisol level on immune responsiveness should be influenced by the associated levels of GH and PRL, to name just two of the many other hormones capable of modulating immune activity. The situation is further complicated by considering peripheral neural activity, which additionally modulates immune responses systemically and within specific tissue microenvironments.

Progress requires human studies that sample multiple hormones, peptides and neurotransmitters, their receptors and serum-binding proteins, and their response to stress over time and in relation to indices of immune responsiveness. In addition, it is necessary to investigate in humans patterns of brain activity using imaging techniques in relation to stress-inducing situations. The brain activity patterns, in turn, must be related to profiles of endocrine and peripheral neural activity. Complexity should be expected in the sense that brain activity patterns will not be linearly correlated with neuroendocrine profiles. Nonetheless, there should be global shifts in how brain activity and neuroendocrine responses are altered by stress when there is also evidence that immune function is impaired.

Clearly, human studies of immune responsiveness can give only a partial picture of the state of the system. Stress has been shown to alter those aspects of immune function that can be monitored. Frequently, the stress effect is initially in the direction of immune mobilization, which then gives way to suppression and, therefore, could increase susceptibility to various disorders that emerge as a consequence of immune dysregulation. However, research that documents immune function, the onset, and course of disease in the same subjects is sorely lacking, as will become apparent when available evidence in this regard is examined in the next two chapters.

Finally, it is clear that neuroendocrine pathways are abundant, but there is not a straightforward mapping of influence. For instance, it is common in studies that have examined both cortisol level and indices of immune responsiveness to find no correlation, although it is well established that cortisol dampens immune reactivity and that stress activates the HPA axis. This is probably an indication that the ultimate action of any given hormone or peptide is highly dependent on the neuroendocrine and genetic context within which it occurs, and therefore, effects do not yield simple linear relationships. However, what seems likely is that *within individuals*, some association should be detectable between the state of different system components (e.g., neural, endocrine, and immune) and risk of illness.

IX. SOURCES

Biondi, M. (2001), "Effects of stress on immune functions: An overview," in Ader, R., Felten, D.L., & Cohen, N. (eds) *Psychoneuroimmunology*, 3rd ed, vol 2, San Diego, Academic Press.

Buckingham, J.C., Cowell, A.M., Gilles, G.E., et al. (1997), "The neuroendocrine system: Anatomy, physiology, and response to stress," in Buckingham, J.C., Gilles, G.E., & Cowell, A.E. (eds) *Stress, Stress Hormones and the Immune System*, Chichester, England, John Wiley & Sons.

Daruna, J.H. (1996), "Neuroticism Predicts Variability in the Number of Circulating Leucocytes," *Personality and Individual Differences*, 20, pp. 103–108.

Dhabhar, F.S. (2002), "Stress-Induced Augmentation of Immune Function—The Role of Stress Hormones, Leukocyte Trafficking, and Cytokines," *Brain, Behavior, and Immunity*, 16, pp. 785–798.

Fehder, W.P., Sacks, J., Uvaydova, M., & Douglas, S.D. (1997), "Substance P: An Immune Modulator of Anxiety," *Neuroimmunomodulation*, 4, pp. 42–48.

Flinn, M.V. (1999), "Family environment, stress, and health during childhood," in Panter-Brick, C. & Worthman, C.M. (eds) *Hormones, Health and Behavior*, Cambridge, England, Cambridge University Press.

Glaser, R. and Kiecolt-Glaser, J. (1994), *Handbook of Human Stress and Immunity*, San Diego, Academic Press.

Goldberger, L. & Breznitz, S. (eds) (1993), *Handbook of Stress: Theoretical and Clinical Aspects*, 2nd ed, New York, Free Press.

Hubbard, J.R. & Workman, E.A. (eds) (1998), *Handbook of Stress Medicine: An Organ System Approach*, Boca Raton, Fla, CRC Press.

McEwen, B.S. & Mendelson, S. (1993), "Effects of stress on the neurochemistry and morphology of the brain: Counterregulation versus damage," in Goldberger, L. & Breznitz, S. (eds) *Handbook of Stress: Theoretical and Clinical Aspects*, 2nd ed, New York, Free Press.

Sapolsky, R.M. (1998), "Hormonal correlates of personality and social contexts: From non-human to human primates," in Panter-Brick, C. & Worthman, C.M. (eds) *Hormones, Health and Behavior*, Cambridge, England, Cambridge University Press.

Ursin, H., Baade, E., & Levine, S. (eds) (1978), *Psychobiology of Stress: A Study of Coping Men*, New York, Academic Press.

Vellucci, S.V. (1997), "The autonomic and behavioural response to stress," in Buckingham, J.C., Gilles, G.E., & Cowell, A.E. (eds) *Stress, Stress Hormones and the Immune System*, Chichester, England, John Wiley & Sons.

Zorilla, E., Luborsky, L., McKay, J.R., et al. (2001), "The Relationship of Depression and Stressors to Immunological Assays: A Meta-analytic Review," *Brain, Behavior, and Immunity*, 15, pp. 199–226.

CHAPTER 8

Infection, Allergy, and Psychosocial Stress

I. **Introduction   134**
II. **Infectious Diseases   134**
   A. Infectious Agents   135
      1. Subcellular agents   136
      2. Unicellular and Multicellular
         Organisms   136
   B. Pathogenic Mechanisms   137
   C. Infection and Other Diseases   138
      1. Cardiovascular Disease   138
      2. Gastrointestinal Tract Disorders   139
   D. Psychosocial Stress, Immunity,
      and Infection   139
      1. Life Events   140
      2. Social Relationships   140
      3. Personality, Behavioral Style,
         and Outlook   141
      4. Immune Mediation of Psychosocial
         Stress Effects   141
III. **Allergic Diseases   143**
   A. Allergens   143
   B. Prevalence and Genetic Influences   144
   C. Environmental Cofactors   145
   D. Pathogenic Mechanisms   145
      1. Food Allergy   146
      2. Parasitic Infection   146
      3. Neurogenic Inflammation   147
   E. Complexity of Allergic Responses   147
   F. Psychosocial Stress, Immunity,
      and Allergy   148
IV. **Concluding Comments   149**
V. **Sources   151**

I. INTRODUCTION

This chapter discusses major disease types in which immune activity plays a role. Specifically, infectious diseases and allergy are considered, both occurring in response to some form of external irritant. The intent is to provide a foundation for understanding the role of psychosocial stress.

As previously noted, disease constitutes stress, but in this chapter, the emphasis shifts to the role of *psychosocial stress*. Evidence that psychosocial stress can increase the probability of disease or exacerbate the manifestation of a disease process is presented. In effect, one type of stress begets another or, in recent terminology, augments allostatic load. Also examined is research indicating that the effect of psychosocial stress on disease may be mediated by modulation of immune function.

II. INFECTIOUS DISEASES

Infectious diseases are by definition transmissible. Outbreaks of infectious diseases have been documented throughout history. One of the earliest recorded epidemics, believed to have been smallpox, occurred in Egypt around 1350 B.C. Other well-documented instances include the plague of Athens around 430 B.C. and the "black death" in Europe from A.D. 1347 to 1351. Epidemics represent dramatic increases in the incidence of cases of a specific malady, possibly reflecting a novel pathogen or novel conditions (e.g., crowding) that facilitate transmission of a pathogen that may have been endemic (i.e., affecting only a stable number of individuals) for some time in the population.

An understanding that such rapidly spreading diseases must be transmitted from individual to individual was evident as early as the sixteenth century, when Francostoro (ca. 1546) theorized that invisible particles (*seminaria*) transmitted infectious diseases. However, it was not until the last quarter of the nineteenth century that specific microorganisms began to be discovered along with the manner in which they were transmitted. For instance, the organism that caused the plague, *Yersinia pestis*, was not identified until 1894. It was found to be transmitted from rats, which served as a reservoir via a vector (rat fleas) that bit rats, humans, and other mammals. This is an example of a zoonotic infection because it is able to cross species. However, its transmission could be made more direct if the infection progressed to the lungs because it then could be passed via the air as a result of being sprayed while coughing.

Despite the devastation caused by the great plague during the Middle Ages, it seemed to gradually disappear. During the London epidemic of 1665, there were approximately 7000 deaths per week in a population of half a million inhabitants, but by 1840 deaths due to the plague had virtually disappeared. It seems likely that rudimentary public health measures coupled with the gradual elimination of highly susceptible individuals may have been largely responsible.

Epidemics, like forest fires, are dynamic processes that require favorable conditions to spread. There must be a large supply of susceptible individuals; the infected individuals must come in contact with susceptibles (or vectors) during their infective period in sufficient numbers so the pool of the infected will expand. The latter depends also on how quickly individuals are removed from the infective pool, via recovery with immunity or death, along with the number of susceptibles being introduced into the population. Thus, there are various ways in which epidemics can recede into an endemic state and even be eradicated from a population.

Infection is not tantamount to disease. First, there is an *incubation* period, which defines the time between infection and the onset of signs or symptoms of disease. *Infectivity* is the ease with which an infection can be transmitted, whereas *pathogenicity* refers to the likelihood that disease will ensue. *Virulence* is a characteristic of the pathogen that predicts the severity of disease and the likelihood of mortality. *Immunogenicity* refers to the fact that organisms differ in their ability to provoke immune responses, although host factors are also at play in this regard. Even when immune responses are evoked, they may be either ineffective at eliminating the pathogen or actually pathogenic by causing injury to healthy tissues. It must also be recognized that infection (i.e., the presence of microorganisms in the body) need not cause disease and may even be protective or only problematic by creating a *carrier state*, in which the host remains healthy but is able to infect others.

Disease caused by an infectious agent may be followed by recovery and long-term immunity. In other cases, the pathogen may remain capable of causing disease recurrences (latent infection) and the infective state or may have caused sequelae (collateral damage) that persist as disability or dysfunction. For instance, a number of autoimmune disorders have an infectious etiology. Furthermore, there are conditions in which the infection and infectivity remain chronic and disease manifestations emerge gradually or increase the probability of malignant disease in the infected tissue. Nearly 15% of all cancers have been linked to chronic infection.

As a population is exposed to an infectious agent and mounts an effective immune response, the probability of an epidemic is diminished. *Herd immunity* is said to have developed, which limits the opportunity for infectives to meet susceptibles, thus curtailing transmission. Public health measures, including vaccination, serve to enhance the natural process of herd immunity.

A. INFECTIOUS AGENTS

The vast majority of microorganisms and infectious particles are not known to cause disease. Infectious agents range in complexity from the acellular to the multicellular.

1. Subcellular Agents

Prions are protein particles without any genetic material that have been implicated in neurological conditions such as kuru, Creutzfeldt–Jakob disease, and Gerstmann–Sträusser–Scheinker syndrome. *Viroids* contain a simple strand of RNA and have not been implicated in human disease. *Viruses* contain genetic information in the form of RNA (retroviruses) or DNA enclosed within a protein coat, which may be further encased by a lipoprotein envelope. Viruses lack the enzymes and protein assembly structures necessary for self-replication. Viruses must rely on the molecular machinery of the cells they infect to replicate. They are "obligate intracellular parasites." Viruses can infect only certain cell types because they must first bind to relatively specific protein structures on the cell surface to gain entry into the cytoplasm. For instance, in the case of human immunodeficiency virus (HIV), the CD4 molecule that is expressed on helper T cells and antigen-presenting cells (APCs) serves as the entryway for the virus, hence its well-known disruption of immune function. Once a virus enters a cell, it becomes disassembled and its genetic material can use the cells' synthesizing machinery to manufacture progeny. The virus components (i.e., nucleic acid sequences and protein coat or capsid) are assembled into a *virion* (the infectious particle) as the concentration of the components increases within the cell. Virions are released either when the cell dies or through a process of exocytosis, in which the virion is further encapsulated by a portion of the infected cell's outer membrane as it is released. Such a lipoprotein envelope enhances the ability of the virion to infect other cells. Finally, viruses can be subclassified based on a variety of characteristics including size, shape, type of nucleic acid, and presence of lipoprotein envelopes.

2. Unicellular and Multicellular Organisms

A variety of unicellular organisms also cause disease. This group is divided into prokaryotes, which lack a discrete nucleus, and eukaryotes, which have well-defined nuclei. The prokaryotes include the *chlamydiae*, which are intracellular pathogens; *mycoplasmata*, the smallest free-living infectious microbe; and *rickettsiae*, which are also intracellular pathogens that must rely on arthropod vectors for transmission.

Bacteria are also prokaryotes. They have been subclassified based on membrane-staining characteristics, shape, and oxygen tolerance. Recognizing that the available schemes are arbitrary, one of the more useful is the Gram stain, which permits bacteria to be subclassified as gram-positive or gram-negative. In addition, bacteria are named based on their shape. Those whose cellular appearance is round or spherical are called *cocci*, those that are rod shaped are named *bacilli*, and those with spiral shapes are referred to as *spirochetes*. Table 8.1 includes examples of

Table 8.1

Bacterial Classification Schemes

Shape	Gram-positive stain	Gram-negative stain
Cocci	*Streptococcus pyogenes* (Pharyngitis/rheumatic fever)	*Neisseria gonorrhoeae* (gonorrhea)
Bacilli	*Listeria monocytogenes* (bacterial meningitis)	*Yersinia enterocolitis* (gastroenteritis)
Spirochete	*Borrelia burgdorferi* (Lyme disease)	*Helicobacter pylori* (gastritis/peptic ulcers)

specific bacteria, illustrating the classification schemes along with the diseases they are known to cause.

Fungi are eukaryotic organisms capable of infection. Fungi come in two forms, unicellular (yeasts) and multicellular (molds). Some exist in both forms (dimorphic) and tend to be high in pathogenicity. Fungi are plantlike and cause infections that are relatively superficial (cutaneous), somewhat deeper (subcutaneous), or systemic, affecting internal organs. For instance, *Candida albicans* is a yeast that can cause infections of mucosal membranes or systemic mycoses. Fungi also cause opportunistic infections (i.e., those that develop only in immunocompromised individuals). A well-known instance of the latter is the pneumonia caused by *Pneumocystis carinii* in patients with acquired immunodeficiency syndrome (AIDS).

Protozoa are another type of eukaryotic organism responsible for severe disease, particularly in underdeveloped regions of the world. Some of the diseases caused by this class of organism include sleeping sickness, malaria, and toxoplasmosis. The latter can be acquired from undercooked meat and is particularly detrimental to the unborn. Many of the protozoan diseases are transmitted via vectors such as the tsetse fly and the anopheles mosquito.

Multicellular organisms classified as *helminths*, which include flatworms and roundworms, are infectious as well. They can infect the intestines and inhabit the liver, lungs, blood, brain, and skin. *Arthropods* such as the scabies mite and the head and body louse are multicellular organisms capable of infecting body surfaces.

B. Pathogenic Mechanisms

The vast diversity of microbial life capable of infection requires that organismic defenses be sufficiently resourceful to dispose of specific invaders. Most microorganisms are not pathogenic or may even be beneficial and, therefore, safely ignored. Those that are pathogenic are so because they have characteristics (e.g., release toxins) that can disrupt normal tissue function or compete with the organism's cells for life-sustaining resources. Some microorganisms cause the death of the cells they infect (e.g., motor neurons in poliomyelitis or T cells in AIDS). Manifestations of illness such as fever and malaise are a consequence of the release of cytokines as leukocytes become activated or degenerate. Damage to tissues can be the result of the immune response itself, as when granulomas develop in response to infection with *Mycobacterium tuberculosis*, in the effort to contain the pathogen or when highly toxic molecules such as tumor necrosis factor -α (TNF-α) cause the destruction of lung tissue. Moreover, a process known as *molecular mimicry* can lead to distal tissue damage. This occurs when antibodies produced in response to a pathogen are able to cross-react with healthy self-tissue and damage it. For instance, the heart tissue damage seen in rheumatic fever is caused by antibodies produced to fight strep-throat infection.

C. INFECTION AND OTHER DISEASES

It is now generally recognized that infection plays a role in disorders such as gastric ulcers and cardiovascular disease. These are diseases that were previously thought to be brought on exclusively by psychosocial stress and associated lifestyles.

1. Cardiovascular Disease

Inflammation is associated with the development of cardiovascular disease, which is a leading cause of morbidity and mortality. Nonspecific markers of systemic inflammation, such as acute-phase proteins like C-reactive protein and fibrinogen, are linked to increased risk of cardiovascular events. Atherosclerotic plaques are not simply an accumulation of lipids. They reflect clonal expansion of cells at the site, due to an excessive healing response after injury of the endothelium. Greaves has described plaque as *minitumors*. They are infiltrated by leukocytes, particularly macrophages and T lymphocytes. The endothelium at the site is characterized by increased expression of cell-adhesion molecules, probably in response to cytokines released in response to local infection.

Several pathogens have been found to increase the risk of cardiovascular disease. Among these are *Helicobacter pylori, Chlamydia pneumoniae,* and *cytomegalovirus* (CMV). Such microorganisms may infect the endothelial cells and become foci for atherosclerotic lesions or they may infect leukocytes in plaque and activate them in a way that enhances lesion progression. Inflammation is a correlate of coronary artery disease, although whether it causes the vascular pathology or simply constitutes an accompanying response to the atherosclerotic disease process is not yet entirely clear.

The growing recognition of immune activity involvement in cardiovascular disease, whether in response to pathogens or to abnormal deposition of lipids in the vasculature, does not negate the importance of other previously recognized risk factors. One must not overlook the fact that increased risk of cardiovascular disease has been linked to low socioeconomic status, chronic hostility, and episodes of fatigue and depression. Acute stress, which evokes increased mental activity and emotional responses, particularly anger, has also been found to precede adverse cardiovascular events. Thus, a variety of factors including infectious agents and inflammatory responses may play a part in causing disorders such as atherosclerosis.

2. Gastrointestinal Tract Disorders

Ewald in his book *Plaque Time* has persuasively argued for the need to recognize the role of pathogens in various chronic diseases. He has relied heavily on the discovery that *H. pylori* infection is associated with peptic ulcer disease and that treatment with antibiotics proves effective, raising doubt about the importance of psychosocial stress, which was previously regarded as a major etiological factor.

Others have noted that *H. pylori* infection is associated with a variety of gastrointestinal (GI) tract disorders. The bacteria do not need to be present in all cases of GI tract disease and are often found in healthy individuals. Moreover, some evidence suggests that gastritis may precede colonization by *H. pylori*. The latter raises the possibility that the association reflects the tendency by the bacteria to colonize epithelial tissues that have become disturbed. In addition, it has been argued that the effectiveness of antibiotics is not adequate proof of causation, because when antibiotics are used alone, they do not perform better than placebo, even though antiacid preparations appear effective alone. The combination of antibiotics with protective coating agents or acid secretion inhibitors is typically needed for improvement.

D. Psychosocial Stress, Immunity, and Infection

The evidence that psychosocial stress is a factor in infectious disease comes from a variety of sources. The focus here is on the human evidence, especially that arising from prospective studies in which psychosocial measures have been collected before the onset of illness. Such studies provide the more compelling evidence for determining which psychosocial factors increase the risk of disease. In addition, consideration is given to studies that evaluate psychosocial variables during or after the diagnostic process, not as indications of risk but as predictors of prognosis. The latter become particularly compelling when coupled with evidence that interventions aimed at the psychosocial variables have beneficial consequences with respect to the course of disease.

1. Life Events

Many studies have found that the amount of life change and the occurrence of distressing events correlate with the onset and severity of clinical symptoms of infection. For instance, young children from families experiencing more life change are more likely to have been hospitalized for treatment of respiratory tract infections. Families experiencing more life change have twice the number of episodes of upper respiratory tract infection and days of illness than those experiencing the least amount of life change over a 6-month period. Young men who are highly motivated to perform well in a military university but who are receiving low grades are more likely to experience severe cases of infectious mononucleosis.

Overall, studies of HIV-infected individuals have failed to show that more life change is associated with the occurrence of physical symptoms indicative of disease progression. Viruses capable of latent infections (e.g., herpes simplex virus) have been found in some studies to be reactivated in response to increased life change. However, this association is not readily found in prospective studies, which are methodologically more rigorous than those that rely on recall of life change during the recurrence of a latent infection (i.e., retrospective studies). In general, across a variety of infectious diseases, there have been enough studies in which life change has been unrelated to the onset or severity of illness to raise doubt concerning the impact of such changes across the board. By the same token, there does not appear to be any reported instance in which more life change has been associated with less illness.

2. Social Relationships

Family atmosphere (i.e., the nature of the relationships among family members) is another variable that has been found to have an impact on illness. Families whose members tend to be excessively distant (detached) or overly involved and intrusive (enmeshed), as well as those families in which routines are inflexible (rigid) or there is much unpredictability and disorganization in day-to-day life (chaotic), are more prone to have cases of influenza. The highest incidence of influenza has been observed in disorganized families.

Social support, which implies some form of interaction with others, has been noted in some studies to increase the risk of clinical symptoms of infection. In fact, more socially integrated individuals who are HIV positive appear at greater risk of progressing to AIDS than those who are more isolated. In effect, social stimulation may increase exposure to other pathogens or have an activating effect on immune function, which in the case of HIV may overwhelm immune defenses or accelerate damage to the immune system and thereby outweigh any benefit of social support.

3. Personality, Behavioral Style, and Outlook

Personality traits and other relatively stable psychological characteristics have been associated with the severity of infections. For instance, introverted individuals appear more prone to exhibit clinical symptoms of infection. A number of studies have explored the role of personality attributes in the outcome of HIV-positive individuals. Hardiness has not been found to be protective. Similarly, generalized measures of hopelessness or depressive outlook are not predictive of clinical decline, although depressive outlook does correlate with T-cell count. However, measures that reflect HIV-specific attitudes predict progression. Fatalistic individuals who have experienced the loss of a partner in the preceding year decline more rapidly. This is also true of individuals who rely on denial to deal with distress. In this regard, drug abusers are also at risk of a more rapid demise. This may not be solely due to drug effects or greater infectious burden; it may to some extent relate to personality attributes that both motivate drug use and accelerate the progression of HIV disease. In effect, more inhibited, passive individuals blame themselves for their plight, have negative expectations regarding the outcome, and tend to progress more rapidly to AIDS. These are attributes that serve to define depression, a condition prospectively found to predict AIDS onset. Individuals who remain chronically depressed are at the greatest risk. Unfortunately, pharmacological treatment for depression does not appear to have much of an impact. HIV-positive individuals who fare better appear to deal with adversity by finding life-affirming meaning and avoiding despair. Their coping style is more active, assertive, and optimistic.

4. Immune Mediation of Psychosocial Stress Effects

Studies that examine measures of stress, immune function, and disease status in the same subjects are becoming more available. These studies are crucial because they may provide evidence that psychosocial stress alters health by how it modulates immune function, especially with respect to infectious diseases, in which case the immune system constitutes the major impediment to rampant growth of the infectious agent and serious illness.

Studies of this type are difficult to conduct in humans because the timing of real-life stress cannot be controlled. Moreover, the relevant facets of immune responsiveness may not be subject to straightforward quantification, and the detection of signs or symptoms of illness may require a level of monitoring that is not generally practical. Ideally, one would like to show in a population that those who have experienced more contextual change recently (i.e., have a greater level of stress) appear initially less immunologically responsive and gradually present with more clinical signs of infectious disease. The current data, thus far, are not entirely consistent with this simple scenario.

As noted earlier in this chapter, a great deal of attention has been given to the role of psychosocial stress in the progression of HIV infection to full-blown AIDS. Some of those studies have quantified psychosocial stress prospectively, have monitored aspects of the immune system in the same subjects (e.g., $CD4^+$ count, lymphocyte proliferative capacity, and natural killer [NK] cell activity), and have assessed the clinical condition of the individuals. Data from such studies indicate that psychosocial stress is related to both immune decline and signs of illness, but whether the individuals who appear more immunologically impaired are the ones showing more signs of disease is not entirely clear. Thus, simple linear relationships linking stress to immune function and then to clinical status have not been observed in this patient population and are probably not the norm.

The studies of Cohen et al. (1997) are the most methodologically sophisticated in this area. They have exercised control over the infectious process by inoculating volunteers with relatively benign viruses that cause upper respiratory tract cold symptoms. The volunteers are then housed under standard conditions for 1 week. They are monitored for the occurrence of infection and the development of cold symptoms. Variation in the occurrence of colds is examined in relation to variables such as (1) the number and duration of burdensome life events, (2) social network diversity (i.e., the variety of social relationships experienced), (3) health habits (e.g., diet, sleep, and substance use), and (4) personality. In an attempt to document pathways whereby psychosocial events affect immune function, these investigators have also measured urinary epinephrine, norepinephrine, and cortisol levels. In addition, they have included measures of immune responsiveness such as NK cell activity and cytokine levels, specifically interleukin-6 (IL-6), which has been found to be increased in relation to the severity of cold symptoms.

Even well-designed studies have significant limitations and can be easily criticized, because it is not possible to adequately address all the pertinent issues in a single study. Nonetheless, these studies have found that stressful life events that tend to persist, the personality dimension of introversion, and low social network diversity are all related to the incidence of colds. Higher levels of norepinephrine and IL-6 are associated with the occurrence of colds. However, neither cortisol level nor NK cell activity predicts the development of cold symptoms. The preceding variables add predictive power independent of type of virus or the titer of antibodies to the specific virus before inoculation. In essence, the psychosocial variables have effects over and above the effects attributable to pathogenicity (i.e., the fact that some viruses caused more colds than others) and the immunological readiness of the individual to deal with a specific virus (i.e., some volunteers had higher specific antibody titers at baseline). These studies are highly informative, even if not entirely clear on whether the effects are first on the immune system and secondarily on the disease process. The IL-6 findings are particularly noteworthy because IL-6 appears to be high in response to both psychological stress and cold symptoms, another form of stress, making IL-6 potentially valuable as a marker of stress level, analogous to how cortisol has been viewed.

III. ALLERGIC DISEASES

The term *allergy* was first used by von Pirquet in 1906 to refer to a variety of clinical disorders, many of which had been recognized in the medical literature for centuries. For instance, asthma had been described by Aretaeus in A.D. 81, and hay fever, so named in 1819 by Bostock, has been recognized since the sixth century A.D. Adverse reactions to food were documented as early as 2500 B.C. What began to be understood at the turn of the twentieth century was that immune mechanisms were responsible for many cases of allergic disease. The term *atopic disease* was proposed by Coca and Cooke in 1923 to denote a subgroup of allergies, which tended to run in families and in which specific immunoglobulin E (IgE) antibodies seemed to be implicated.

An essential aspect of allergic disorders is inflammation that can be more or less localized and that occurs in response to relatively innocuous external irritants (i.e., allergens). Inflammation of the respiratory tract mucosa results in conditions such as *rhinitis*, commonly known as *hay fever*, and *asthma*, when the bronchi are affected. Inflammation of the skin can present as *atopic dermatitis*, also known as *eczema*, *urticaria* or hives, and *angioedema* when the deeper layers of the skin are affected. The skin is a common site of allergic responses even when substances do not make contact with the skin. Ingestion of particular substances or drugs can cause reactions on the skin. *Anaphylaxis* is a particularly dangerous systemic inflammatory response. There may be an initial feeling of "impending doom," feeling as if something is wrong (in other words, the brain is sensing unexpected changes in the body). This feeling may lead to sensations of tingling, warmth, and tightness. If the process is not halted quickly, it may become difficult to swallow and itching will ensue, along with abdominal cramps and diarrhea. Finally, hypotension, upper respiratory tract obstruction, and cardiorespiratory arrest can lead to death.

A. ALLERGENS

Allergens are typically proteins arising from plant life or organisms and may be inhaled, ingested, or absorbed through the skin. Pollen and dust mites give rise to airborne allergens often responsible for allergic rhinitis. Most food allergies have been attributed to protein components of cow's milk, chicken eggs, peanuts, soybeans, cereals, nuts, and marine organisms. Beef and chicken rarely cause allergic responses. Insect bites and contact with certain plants (e.g., poison ivy) can trigger allergic responses. Pharmaceuticals can also function as allergens when applied to the skin or ingested. They may contain protein material that evokes an allergic response or may combine with endogenous proteins (i.e., function as a hapten) to trigger an allergic response. Drugs can cause rashes and serious inflammation of the skin and mucosal membranes (e.g., Stevens–Johnson syndrome).

B. Prevalence and Genetic Influences

Approximately 17% of the general population suffers from allergies. The vast majority is afflicted with hay fever or asthma. These conditions account for nearly two–thirds of the cases of allergic disease. Atopic dermatitis affects at least 3% and food allergies afflict approximately 2% of the general population. However, there is evidence that allergic disease is on the increase.

Susceptibility to allergy is partially rooted in the individual's genetic makeup. Allergy tends to run in families; 40–80% of allergic individuals have relatives with allergies, whereas the figure for nonallergic individuals is closer to 20%. Concordance for asthma and atopic dermatitis is much higher in monozygotic than dizygotic twins. Concordances as high as 70% have been reported for monozygotic twins in contrast to 20% for dizygotic twins. However, the most extensive study of allergy in twins suggests a less marked difference between monozygotic and dizygotic twins. These data were derived from retrospective self-reports, but the sample was quite large, consisting of approximately 7000 twin pairs (approximately 3000 males and 4000 females). Asthma was reported by 3.8%, hay fever by 14.8%, and eczema by 2.5% of the sample. At least one of those disorders afflicted 18% of the sample. The concordances for monozygotic and dizygotic twin pairs are shown in Table 8.2.

The genetic contribution appears highest for asthma and eczema but is modest overall. Also evident is that having one allergic disorder tends to increase the probability that one will have another condition (e.g., asthma increases the

Table 8.2

Concordance of Prevalent Allergic Diseases

Disorder	Concordances	
	Monozygotic	Dizygotic
Asthma	19.0%	4.8%
Hay fever	21.4%	13.6%
Eczema	15.4%	4.5%
At least one of the above	25.3%	16.2%

likelihood of hay fever or eczema). This study further indicates that allergic individuals are at increased risk for disorders not generally regarded as "allergic" such as headaches and tachycardia. Moreover, particularly noteworthy is that allergic individuals have a higher prevalence of autoimmune disorders such as rheumatic fever and rheumatoid arthritis.

Genes regulating IgE production may be involved in increasing the risk of allergy. Individuals susceptible to allergy have three to five times higher IgE levels than nonatopic individuals. In addition, genes regulating IgE receptor density on mast cells, release of inflammatory mediators, and end-organ responsiveness are likely to contribute to risk. A gene that regulates Th1/Th2 balance may be involved as well.

C. Environmental Cofactors

Allergic responses are influenced by environmental factors beyond exposure to allergens. This is suggested by the increasing incidence of allergic disease. Possible explanations could include greater exposure to human-made irritants. Irritants such as air pollution and tobacco smoke are often suspected. Ultraviolet radiation and even air temperature could influence the skin and mucosal tissues in a direction that could make them more susceptible to inflammatory responses. By inducing inflammation on their own, infections could alter the impact of allergens, and the epidemiological data suggest that in populations in which infection is less of a burden, allergy tends to be more prevalent. This is especially true when individuals are less susceptible to parasitic infections. However, data have implicated more frequent exposure to antibiotic treatment during the first year of life as a better predictor of subsequent allergy than episodes of infection alone.

D. Pathogenic Mechanisms

Allergens must be processed by the immune system before they are able to trigger allergic responses. This begins with the allergen acting as antigen presented by macrophages or dendritic cells to helper T cells of the Th2 variety. The latter stimulate specific B cells to produce IgE capable of binding to the allergen. The IgE then coats mast cells in specific tissues. Reencountering the particular antigen causes the mast cell to release substances that promote inflammation. These include preformed substances like histamine and other proinflammatory mediators such as the products of arachidonic acid metabolism, which include the prostaglandins and the leukotrienes. There is an immediate permeability of small blood vessels, leading to edema and movement of leukocytes, such as the eosinophils, capable of further amplifying and sustaining the inflammatory response. The recruitment of basophils, eosinophils, macrophages, and platelets into the region can cause a chronic state of inflammation. The mast cell response can be local or progressively widespread to the extent of becoming systemic.

1. Food Allergy

A closer look at the development of food allergy is instructive with respect to mechanisms underlying allergy. Food allergens can have effects outside the GI tract and can even trigger anaphylaxis. The GI tract is a major site where the organism encounters a wide array of proteins arising not only from food but also from pathogens. Therefore, it is endowed with layers of protection including enzymes that break up proteins and secretions, which can wash away potentially offensive particles. The GI tract also has a vast array of immunological defenses capable of processing proteins typically encountered in food and developing tolerance for them. This is an active process that depends on resident $CD8^+$ T cells. However, if the GI tract lacks a full complement of protective mechanisms, food allergies may develop. This can occur through exposure to food substances such as cow's milk during early infancy when the GI defensive repertoire is not yet fully developed. In most individuals, specific food allergies tend to be outgrown (the figure may be as high as 87%), but in some, such allergies persist and may expand to include other foods.

Allergic responses to food are IgE mediated and reflect mast cell degranulation in the GI tract or at distal sites that have been sensitized. Such distal sensitization requires the entry of high enough levels of IgE into distal tissues or the local presence of allergen-specific plasma cells capable of producing the IgE. Their activation, in turn, would have resulted from the arrival of specific Th2 cells able to stimulate resident B cells to produce specific IgE. In essence, the pattern of an allergic response to food depends on the distribution of specific IgE mounted on mast cells and the dose of allergen arriving at the sensitized location. The response can range from mild and highly localized inflammation to severe systemic life-threatening inflammation. Disorders as diverse as arthritis and migraine, even attention-deficit/hyperactivity disorder, have been attributed to food allergy. With respect to the latter, it should be noted that mast cells are found within the central nervous system (CNS).

2. Parasitic Infection

Allergic disease is predominantly mediated by humoral immunity and is reliant on IgE. It is unlikely that the IgE response evolved as a mechanism for promoting allergy. Actually, the IgE response is an effective method of defending against infection by multicellular parasites. In fact, parasitic infections seem to decrease the probability of allergic disease, as if when the IgE response is engaged by actual threat, it is less likely to be triggered by allergens. Thus, one way to conceptualize allergic susceptibility is as a robust ability to respond to parasites, which in their absence can be triggered by innocuous molecular structures, perhaps because of some homology with parasitic antigens. The latter is simply a hypothesis, but one that is consistent with evolutionary principles.

3. Neurogenic Inflammation

A most important observation, in the case of allergic disease, is that sensory nerve endings in the area of inflammation respond to the proinflammatory mediators released by mast cells. This response is the source of nociceptive signals sent to higher structures within the CNS. The response, via a local axon reflex, also leads to the release of peptides in the area, which can further amplify inflammation.

The localization of such proinflammatory peptides is not restricted to sensory nerve endings. Tachykinins such as substance P and neurokinin A, as well as corticotropin-releasing hormone, are found in the autonomic nervous system and are capable of mediating neurogenic inflammation either in response to local activation by axon reflexes or due to centrally triggered release. Parasympathetic activity appears capable of causing tissue-specific or regionally restricted inflammation. The importance of neurogenic inflammation cannot be overemphasized because it constitutes a mechanism whereby mental phenomena can affect the condition of peripheral tissues and may even promote disease.

E. COMPLEXITY OF ALLERGIC RESPONSES

The complexity of allergic phenomena is nicely illustrated by airway hyperreactivity, which results from edema, mucous hypersecretion, and bronchoconstriction. Such local tissue changes occur in response to mast cell degranulation and in response to the influence of nociceptive neurons, parasympathetic terminals, and epinephrine released by the adrenal medulla.

The major pathway leading to mast cell degranulation requires that allergen be processed as antigen by APCs presented to Th2 cells, which then promote the synthesis and release of allergen-specific IgE. The latter, in turn, sensitizes mast cells so future encounters with allergen will cause mast cell degranulation and the allergic response. Mast cell degranulation and local tissue responses can be facilitated by the chemical composition of the local microenvironment.

Peptides released from nociceptive sensory fibers in the bronchi promote constriction, the degree of which depends on the levels of neutralizing enzymes available in the region. Cholinergic input further increases bronchoconstriction, an effect that is counteracted by vasoactive intestinal peptide and nitric oxide, both co-released with acetylcholine from parasympathetic terminals. Epinephrine from the adrenal medulla also acts to counteract bronchoconstriction. Clearly, the net effect will reflect relative concentrations of these various substances, as well as local receptor densities.

However, neural pathways are not crucial to allergen-induced bronchoconstriction. Neural signals operate in parallel to allergen-provoked signals, although at some level, probably intracellular, they are likely to interact and thereby initiate a response. Manipulation of peptides and neurotransmitters in normal individuals is not sufficient to provoke asthma. In asthmatics, bronchoconstriction can be induced by placebo and more generally in response to suggestion. The placebo effect appears to be vagally mediated, as it can be blocked by anticholinergic drugs.

Finally, the treatments of choice to block bronchoconstriction (i.e., glucocorticoids and β-adrenergic agonists) tend to suppress cell-mediated immunity (Th1) and augment humoral immunity (Th2), so although in the short-term they are beneficial, in the longer term they may aggravate the underlying potential for severe bronchoconstriction, by causing an increased production of IgE.

The picture that emerges is one of individuals with a robust IgE response who may overreact to innocuous substances when spared confrontation with parasites. The immune response aimed at suspected pathogens that are not present will cause injury and dysfunction of healthy tissue. Moreover, available data clearly show that the neural circuitry in the region can modulate the inflammatory process. In some instances, neural signals may amplify inflammation and in rare cases may even initiate a full-blown allergic response. Finally, treatments that have a clear immediate benefit may at the same time aggravate the underlying condition.

F. PSYCHOSOCIAL STRESS, IMMUNITY, AND ALLERGY

Allergic disorders such as asthma and atopic dermatitis were regarded as classic psychosomatic disorders in the heyday of psychosomatic medicine from the 1930s to the 1950s. In fact, asthma was thought to be primarily a nervous disorder (asthma nervosa) attributed to disturbances in the mother–child relationship. This will sound familiar to anyone conversant with early psychodynamic explanations for disorders such as autism, schizophrenia, and anorexia nervosa, which were not well understood and affected relationships within the family.

Asthma is now typically regarded as an allergic disorder that can also be triggered by viral infections and even strenuous physical activity in some individuals. Many experts do not seem to assign much weight to psychosocial factors with respect to the onset or course of allergic disorders. In fact, an interesting observation regarding the clinical manifestation of asthma is that under some circumstances, psychosocial stress may prove beneficial. For instance, one early study by

Funkenstein (1953) found that if one provoked an asthmatic response by administering a drug (Mecholyl) that precipitates asthma in asthmatics only, the response was less pronounced if the individual was under stress (e.g., the night before final examinations) at the time of the challenge.

Given the possible beneficial effect of psychosocial stress, it is not surprising that the role of life change in modulating allergic disorders has been less clear than in the case of infectious disease. Similarly, a propensity toward negative affectivity has been observed only in some samples of allergic individuals, but even then, whether negative affectivity is a contributing factor or simply a consequence of the allergic condition is not clear.

Family dynamics, in particular the presence of dysfunctional patterns of interaction, has been found to affect the prevalence of asthma. This effect seems most pronounced when there is no family history or other evidence of increased susceptibility to asthma. For instance, in a prospective study, Faleide, Unger, and Watten (1987) reported the prevalence of asthma to be higher, irrespective of evidence of susceptibility (e.g., increased IgE), in dysfunctional families compared with well-functioning families. The family dysfunction effect was most dramatic in individuals whose genetic susceptibility appeared to be minimal. However, other research suggests that family dysfunction may develop as a result of having an asthmatic child. An additional question in this regard concerns whether the effect of family dysfunction reflects the socioemotional climate in the home or the allergen burden due to the physical conditions within the home.

An especially important discovery is the observation that allergic responses can be conditioned in both animals and humans. In other words, the allergic response can be triggered by settings or procedures that have been associated with allergen, even when allergen is not administered. Moreover, such conditioned allergic responses can be extinguished, as is expected of any learned association. Thus, the fact that allergic responses can be conditioned makes such responses susceptible to the influence of psychosocial factors.

IV. CONCLUDING COMMENTS

All diseases are forms of stress in the sense that they constitute change in the intraorganismic context. In the case of infectious diseases, the intraorganismic change occurs as a result of invasion by diverse microorganisms. Microorganisms capable of causing human disease are only temporarily detrimental, that is, until evolutionary processes within the individual lead to adaptation. The latter occurs through the action of the immune system, which renders a sector of the human population better equipped to dispose of the specific pathogen. At another level, through the capability of the nervous system to detect order, individuals learn to avoid exposure to specific pathogens by instituting public health measures, which reduce exposure and block pathogen transmission. Infection, viewed as stress, serves to highlight that stress has detrimental and survival-enhancing effects, at least at the population level. In other words, although contextual change will be associated with casualties, it will also augment the adaptation of the surviving population to the new context.

Psychosocial stress represents another facet of contextual change that may increase the probability of disease or enhance its progression and thereby produce additional stress. The available data lend some support to the role of psychosocial stress in infectious disease. Life change is at least a weak predictor of the onset of infectious disease. Studies focusing on family dynamics suggest that disturbed relationships predict increased episodes of infectious disease. Data on social support more generally suggest that the effect may depend on other variables including type of infection. Personality characteristics such as introversion and chronic states of negative emotion such as depression are associated with a higher probability of clinical symptoms of infection and a more rapid progression to disease in the case of HIV infection. Well-controlled studies of upper respiratory tract infections further support the role of psychosocial variables as predictors of the response to experimentally induced viral infection.

It is clear that diseases interact in complex ways. Infectious diseases create conditions that can lead to various other chronic diseases. The role of infection is now recognized in cardiovascular disease, peptic ulcers, and even psychiatric disorders. However, recognition of the role of pathogens in such diseases should not eclipse awareness of the influence of psychosocial stress. Psychosocial stress is a factor whose contribution may be obscured because it is not always necessary or sufficient to cause disease. Moreover, psychosocial stress may have diametrically opposed effects across individuals and diseases.

Allergy is essentially inflammation triggered by mast cell degranulation in response to relatively innocuous external irritants. Mast cells are widely distributed throughout the body and have been demonstrated in the brain. Nerve endings in the vicinity of mast cells respond to degranulation, via local reflexes capable of amplifying the inflammatory process, and cause signals to travel to the CNS. The possibility for CNS-initiated inflammation is suggested by such innervation and is supported by evidence that in asthmatics, bronchoconstriction can be induced by placebo and more generally modulated in response to suggestion.

The incidence of allergic disease appears to be increasing. It seems that individuals with fewer episodes of infection, particularly parasitic infections, and more frequent use of antibiotics during early life tend to be at higher risk of allergy. An additional point of interest is that treatments such as glucocorticoids, which prevent symptoms of allergy, can simultaneously enhance IgE production and thereby aggravate the underlying problem.

Allergic disorders were once regarded as primarily psychosomatic conditions. Nowadays, experts do not emphasize psychosocial stress as a factor of significance. There is some acknowledgment that family dynamics may play a role in asthma, although data have been interpreted as indicating that asthma causes family dysfunction. Especially noteworthy is the evidence that psychosocial stress may prevent asthmatic responses under some circumstances and that allergic responses can be triggered by neutral stimuli that have been previously associated with allergen.

V. SOURCES

Alexander, F. & Flagg, G.W. (1965), "The psychosomatic approach," in Wolman, B. (ed.) *Handbook of Clinical Psychology*, New York, McGraw-Hill.

Castes, M., Hagel, I., Palenque, M., et al. (1999), "Immunological Changes Associated with Clinical Improvement of Asthmatic Children Subjected to Psychosocial Intervention," *Brain, Behavior, and Immunity*, 13, pp. 1–13.

Cohen, S., Doyle, W.J., Skoner, D.P., et al. (1997), "Social Ties and Susceptibility to the Common Cold," *Journal of the American Medical Association*, 277, pp. 1940–1944.

Cole, S.W. & Kemeny, M.E. (2001), "Psychological influences on the progression of HIV infection," in Ader, R., Felten, D.L., & Cohen, N. (eds) *Psychoneuroimmunology*, 3rd ed, vol 2, San Diego, Academic Press.

Cookson, W.O.C.M. & Moffatt, M.F. (1997), "Asthma: An Epidemic in the Absence of Infection?" *Science*, 275, pp. 41–42.

De Aranjo, G., Van Arsdel, P.P, Jr, Holmes, T.H., & Dudley, D.L. (1973), "Life Change, Coping Ability and Chronic Intrinsic Asthma," *Journal of Psychosomatic Research*, 17, pp. 359–363.

Edfors-Lubs, M-L. (1971), "Allergy in 7000 Twin Pairs," *Acta Allergologica*, 26, pp. 249–285.

Elenkov, I.J. & Chrousos, G.P. (1999), "Stress Hormones, Th1/Th2 Patterns, Pro/Anti-inflammatory Cytokines and Susceptibility to Disease," *Trends in Endocrinology and Metabolism*, 10, pp. 359–368.

Ewald, P.W. (2000), *Plague Time: How Stealth Infections Cause Cancer, Heart Disease, and Other Deadly Ailments*, New York, Free Press.

Fahdi, I.E., Gaddam, V., Garza, L., et al. (2003), "Inflammation, Infection, and Atherosclerosis," *Brain, Behavior, and Immunity*, 17, pp. 238–244.

Faleide, A.O., Unger, S., & Watten, R.G. (1987), "Psychosocial factors in bronchial asthma and allergy in childhood: A prospective study," in Christodoulou, G.N. (ed) *Psychosomatic Medicine: Past and Future*, New York, Plenum Press.

Feaster, D.J., Goodkin, K., Blaney, N.T., et al. (2000), "Longitudinal psychoneuroimmunologic relationship in the natural history of HIV-1 infection: The stressor-support-coping model," in Goodkin, K. & Visser, A.P. (eds) *Psychoneuroimmunology: Stress, Mental Disorders, and Health*, Washington, DC, American Psychiatric Press.

Friedman, H.S. & Booth-Kewley, S. (1987), "The 'Disease-Prone-Personality': A Meta-analytic View of the Construct," *American Psychologist*, 42, pp. 539–555.

Funkenstein, D.H. (1953), "The Relationship of Experimentally Produced Asthmatic Attack to Certain Acute Life Stresses," *Journal of Allergy*, 24, pp. 11–17.

Gustafsson, P.A., Bjorksten, G., & Kjellman, N.I. (1994), "Family Dysfunction in Asthma: A Prospective Study of Illness Development," *Journal of Pediatrics*, 125, pp. 493–498.

Hoeprich, P.D., Jordan, M.C., & Ronald, A.R. (eds) (1994), *Infectious Diseases: A Treatise of Infectious Processes*, 5th ed, Philadelphia, JB Lippincott Co.

Irwin, M. (2002), "Psychoneuroimmunology of Depression: Clinical Implications," *Brain, Behavior, and Immunity*, 16, pp. 1–16.

Kemeny, M.E. (1994), "Stressful events, psychological responses and progression of HIV infection," in Glaser, R. & Kiecolt-Glaser, J. (eds) *Handbook of Human Stress and Immunity*, San Diego, Academic Press.

Kop, W.J. & Cohen, N. (2001), "Psychosocial risk factors and immune system involvement in cardiovascular disease," in Ader, R., Felten, D.L., & Cohen, N. (eds) *Psychoneuroimmunology*, 3rd ed, vol 2, San Diego, Academic Press.

McKeever, T.M., Lewis, S.A., Smith C., et al. (2002), "Early Exposure to Infections and Antibiotics and the Incidence of Allergic Disease: A Birth Cohort Study with the West Midlands General Practice Research Database," *Journal of Allergy and Clinical Immunology*, 109, pp. 43–50.

Meijer, A.M., Griffioen, R.W., van Nierop, J.C., et al. (1995), "Intractable or Uncontrollable Asthma: Psychosocial Factors," *Journal of Asthma*, 32, pp. 265–274.

Nelson, K.E., Williams, C.M., & Graham, N.M.H. (eds) (2001), *Infectious Disease Epidemiology: Theory and Practice*, Gaithersburg, Md, Aspen Publishers.

Padur, J.S., Rapoff, M.A., Houston, B.K., et al. (1995), "Psychosocial Adjustment and the Role of Functional Status for Children with Asthma," *Journal of Asthma*, 32, pp. 345–353.

Schneiderman, N., Antoni, M., Ironson, G., et al. (1994), "HIV-1, immunity, and behavior," in Glaser, R. & Kilcolt-Glaser, J. (eds) *Handbook of Human Stress and Immunity*, San Diego, Academic Press.

Spahn, J.D. & Szefler, S.J. (2002), "Childhood Asthma: New Insights into Management," *Journal of Allergy and Clinical Immunology*, 109, pp. 3–13.

Theoharides, T.C. (2002), "Mast Cells and Stress—A Psychoneuroimmunological Perspective," *Journal of Clinical Psychopharmacology*, 22, pp. 103–107.

Vogel, G. (2002), "Missing Gene Takes Mice's Breath Away," *Science*, 295, p. 253.

Weiner, H. & Shapiro, A.P. (2001), "Helicobacter pylori, immune function, and gastric lesions," in Ader, R., Felten, D.L., Cohen, N. (eds) *Psychoneuroimmunology*, 3rd ed, vol 2, San Deigo, Academic Press.

Wjst, M., Roell, G., Dold, S., et al. (1996), "Psychosocial Characteristics of Asthma," *Journal of Clinical Epidemiology*, 49, pp. 461–466.

Zorrilla, E.P., McKay, J.R., Luborsky, L., & Schmidt, K. (1996), "Relation of Stressors and Depressive Symptoms to Clinical Progression of Viral Illness," *American Journal of Psychiatry*, 153, pp. 626–635.

CHAPTER 9

Cancer, Autoimmunity, and Psychosocial Stress

I. **Introduction   154**

II. **Cancer   154**
   A. Cancer as an Expanding Clone   155
      1. Cancer Cells Evolve   156
      2. Cancer-promoting Genes   156
   B. Environmental Carcinogenesis   157
      1. Toxins   157
      2. Diet and Hormones   158
      3. Infection   158
   C. Defenses against Cancer   158
   D. Psychosocial Stress, Immunity,
      and Cancer   159
      1. Depression, Psychosis,
         and Outlook   160
      2. Social Relationships   161
      3. Life Events   161
      4. Psychosocial Treatment
         and Cancer Survival   161
      5. Immune Mediation
         of Psychosocial Effects   162
   E. Could Psychosocial Factors
      be Irrelevant?   163

III. **Autoimmune Diseases   164**
   A. Clonal Selection Theory
      and Normal Autoimmunity   164
   B. Prevalence of Autoimmune Disorders   165
   C. Pathogenic Mechanisms   165
      1. Autoantibodies   165
      2. Self-reactive T Cells   166
      3. Costimulation   166
      4. Immune Complexes   166
   D. Infection Triggers Autoimmunity   167
   E. Cancer Triggers Autoimmunity   168
   F. Toxic Chemicals
      Trigger Autoimmunity   168

G. Susceptibility to Autoimmunity   169
H. Psychosocial Stress
and Autoimmune Disorder   170
1. Thyroid Disease   170
2. Rheumatoid Arthritis   170
a. Personality and Relationships   171
b. Genes and Circumstances   171
c. Immune Mediation   172
d. Pregnancy   172
3. Insulin-dependent Diabetes   173
4. Inflammatory Bowel Disease   173
5. Systemic Lupus Erythematosus   173
6. Multiple Sclerosis   174
I. Gender and Autoimmunity   174
IV. **Concluding Comments 175**
V. **Sources 177**

I. INTRODUCTION

The previous chapter focused on diseases for which external irritants such as pathogens and allergens play a crucial role. In contrast, this chapter examines diseases that arise when self-tissues become malignant or when immune activity is directed at healthy tissue. Although infectious agents can be a factor in these disorders as well, they are not always implicated. Genetic factors play a role in creating susceptibility to cancer and autoimmune disorders. Psychosocial stress constitutes an additional burden on the organism that may shift the balance so malignancy or autoimmune disorders become more or less likely.

II. CANCER

In contrast to infectious disease, in which external life forms pose a threat to health, in the case of cancer or neoplastic disease, the threat arises from within. The individual's own cells undergo transformation so they multiply uncontrollably and compete with healthy tissue for both space and resources.

Cancer has afflicted humans since prehistoric times. Archaeological evidence places cancer as a malady that afflicted *Australopithecus* and *Homo erectus* well over one million years ago. As such, it predates any of the modern insults that have often been regarded as etiologic. Tumors have been observed in the Egyptian mummies dating to 5000 years ago. The Greeks are credited with recognizing cancer as a distinct disease. Hindu and Chinese medical texts referred to various tumors well over 2500 years ago. Hippocrates and later Galen are credited with the idea that excess of black bile (one of the basic humors), which caused melancholia, also caused cancer.

Nineteenth century physicians were impressed with the rarity of cancer in less developed societies and regarded cancer as a disease of civilization and the associated stressful lifestyles. This general trend is evident in modern statistics, which show that developing countries, where 75% of the world's population resides, account for only 50% of cancer mortality. The other 25% of the global population, residing in the technologically advanced nations, contribute disproportionality to cancer mortality. This is most likely not due to a greater burden of such advanced societies, but to the improved health (longevity) and abundance (dietary and other excesses) of more affluent societies.

Recognition that cancer starts as a small local lesion has been attributed to a French surgeon, Henri Francois Le Dron, who published this idea in 1757. By the mid-nineteenth century, Wilhelm Waldeyer had laid much of the modern foundation for our current views. He suggested that cancer arose by the transformation of normal cells into malignant ones that were capable of movement via the blood or the lymphatic system. It was not until the 1960s that Brookes and Lawley discovered that carcinogenic chemicals damaged DNA, as was already known to be the case for ionizing radiation.

A. CANCER AS AN EXPANDING CLONE

It is now understood that a cancer is an expanding clone. It begins when a single cell becomes transformed, through the accumulation of specific genetic mutations, so that it reproduces frequently and remains insensitive to signals for growth inhibition. This phenomenon is possible because cells have an inherent ability to multiply (this is what gives rise to the organism in the first place), but it is normally regulated to maintain organismic integrity. Under some conditions, a specific cell in a given tissue can become stuck in this growth mode so that only the demise of the organism becomes effective in stopping the runaway proliferation of the immortalized clone arising from the original aberrant cell.

The fundamental mechanisms underlying the original malignantly transformed cell is damage to DNA segments containing specific genes involved in DNA repair, cell growth, and cell death. DNA repair is crucial because if it is faulty, there is an increased probability that genetic mutations affecting cell growth or death will occur. Mutations that augment cells' ability to proliferate or diminish its responsiveness to stop growth signals, including the signals that activate cell death (apoptosis), are fundamental to the initiation of cancer. Genes that when damaged promote growth have been labeled *oncogenes*, whereas those that when damaged preclude stop-growth or cell-death signals from being effective have been designated *tumor suppressor genes*. This is an unfortunate terminology, because the genes are not there to cause or suppress tumors. It is only when genes cease to function normally because of damage (mutations) or some other form of inactivation (e.g., methylation) that they create conditions that can lead to cancer if enough mutations have accumulated in the right genes within a single cell. Moreover, the cell must survive possible destruction by the immune system. At this point, the state of the immune system can become a factor in some forms of cancer.

1. Cancer Cells Evolve

In an informative book, Greaves makes the case that cancer evolves the same way that species evolve. Contextual factors create the conditions in which cancer cells arise and must survive: Those that are best suited and least affected by the defenses of other cells and tissues gain prominence. Therapeutic interventions may temporarily halt progression, but at the same time, they constitute the type of challenge that evolution is able to surmount. This is much like what is now well recognized in the case of pathogens evolving around drugs designed to kill them.

2. Cancer-promoting Genes

The genes that appear mutated in cancer cells are typically those that are involved in some aspect of cell growth (e.g., *ras*, *myc*, and *neu* genes) or cell death (e.g., *p53*). The *p53* gene is believed to be the most commonly mutated gene in human cancer. The p53 protein functions in the activation of cell death in response to DNA damage. It serves to eliminate mutated cells, and when p53 is not functioning, such cells accumulate and essentially become a breeding ground for cancer. Typically, it takes more than one mutation to unleash a cancer clone. The number may be as high as 10 mutations to key genes, although in some tissues during early life the number may be much lower. The need to accumulate mutations is probably the major reason that cancer is much more likely with advanced age. Cancers of infancy and childhood tend to occur in tissues (e.g., nervous system and lymphoid organs) that are undergoing proliferation as part of normal development (e.g., neuroglia) or that are poised to proliferate as part of normal function (e.g., leukocytes). In general, tissues that are self-renewing (e.g., epithelium) are usually the site of cancers throughout the life span.

A proclivity to cancer can be inherited in a variety of ways. Typically, a defective gene is passed on from one parent. In such cases, there is an increased incidence of the particular cancer within the family and the inherited genetic abnormality will be found in all the individual's cells, because it was present in the fertilized egg. Somatic mutations (i.e., those that arise after the organism has been formed) appear only in the cells of the affected tissue. Most cancers arise due to somatic mutations, and even when a genetic defect has been inherited, additional mutations will have to accrue before a cancerous cell emerges. Researchers have documented about 200 genetic defects within human cancer cells. As noted earlier, it may be necessary for at least 10 defects in a given cell to occur before it will be transformed. The odds appear to be against the convergence of mutations in a particular cell, but given enough cells undergoing division and enough time, it becomes inevitable. This is certainly supported by the observation that in the United States, cancer accounts for 23% of all deaths.

B. ENVIRONMENTAL CARCINOGENESIS

It is now well accepted that the genetic damage leading to cancer occurs as a result of a variety of insults that can be traced in part to cultural factors and technological developments that characterize a particular society. The most compelling evidence for this point of view is the observation that there is significant variation in the occurrence of specific cancers across the world. Moreover, when groups of individuals living in one part of the world migrate to another country, they begin to show the cancer-risk profile of their new country. Clearly, their genetic susceptibilities have not changed, so the answer must be sought in the impact of the environment and the associated lifestyles.

1. Toxins

Looking at the environment, one finds several sources of insult. Often these agents are referred to as *carcinogens*. They can arise as a result of "human-made" activities (e.g., manufacturing processes) that generate toxic substances, even those intended to be pleasurable (tobacco smoking). Indeed, tobacco use emerges as the single most destructive source of exposure in modern society, accounting for 30% of all cancer-related deaths. It is ironic that early claims for tobacco benefits included protection against cancer. Other sources of insult to DNA include radiation, both solar and ionizing (i.e., radioactivity), dietary practices, and exposure to infectious agents. With respect to exposure to radioactivity, it is worth noting that the most intense punctuate exposure in the history of humanity occurred after the atomic bomb explosions in Japan. Those who survived the destructive power of the explosion were at increased risk of cancer, particularly leukemia, but in fact, the vast majority of those exposed to high levels of radioactivity did not develop cancer.

2. Diet and Hormones

Diet has become a factor in the sense that our intake greatly exceeds our need for energy, given our sedentary lifestyles. The storage of excess calories in the form of fat and the increased generation of oxygen-free radicals from the induced metabolic activity add to the threat of DNA damage. The endogenous generation of oxygen-free radicals is a major source of DNA damage. Excessive food intake may also mobilize endocrine activity in ways that could enhance the proliferation of cancerous cells. In the case of cancers in reproductive organs, particularly in females, alterations in the exposure of tissues to hormonal signals may facilitate malignancy. For instance, it has long been known that women, such as nuns, who experience more menstrual cycles, as a result of not having children, have been more susceptible to breast cancer.

3. Infection

An estimated 15% of all cancers may result from persistent infection with common viruses or other pathogens. In most cases, infection is not sufficient for the cancer to emerge. Viruses, in particular, can become integrated into the individual's DNA, and depending on where the viral genes introduce themselves into the chromosomes, disruption of crucial genes may or may not occur. Even then, it may be necessary for additional mutations to accumulate before malignancy occurs. Epstein–Barr virus (EBV) is the only virus known to acutely cause malignancy (Burkitt's lymphoma) in humans.

Infectious agents can lead to cancer in a variety of ways. For instance, in the case of hepatocellular carcinoma associated with hepatitis B virus, the mechanism involves the destruction of the hepatocytes by the virus and cytotoxic T cells, which leads to hepatocyte regeneration. The latter process creates conditions for mutations to occur, which along with the assistance of carcinogens can cause malignancy. In effect, immune responses to chronic infection and tissue regeneration to sustain function can increase the probability that the specific tissue will become cancerous.

C. DEFENSES AGAINST CANCER

The primary defense against the occurrence of cancer appears to be the activation of cell death when DNA repair has proven ineffective. Once a cancer clone begins to develop, the immune system could play a role. Tumor-specific antigens could serve to signal the immune system of aberrant tissue change. Such tumor antigens may result from the activation of previously silent genes, mutated genes, or viral genes, in the case in which oncogenic viruses are at work.

Other alterations in the cells' protein products or quantitative changes in the expression of particular proteins can create antigenicity. Both tumor-specific cytotoxic T cells and antitumor antibodies have been documented and could clearly play a role. Consistent with this is the observation that tumors are infiltrated by leukocytes along with the well-documented finding that immunosuppressed individuals have an increased incidence of cancer, largely due to lymphoid tissue cancers.

Finally, the role of immune defenses in controlling malignancies has been very actively explored with respect to cancer treatment. A variety of immunotherapies have been attempted. Essentially, these have been efforts to potentiate the immune activity against specific tumors. Occasionally, these interventions have been quite effective, but overall their impact has been to increase survival for a modest period in a small percentage of patients. For a historical account of such efforts, see Hall's book (1997) *A Commotion in the Blood.*

D. Psychosocial Stress, Immunity, and Cancer

Cancer was a familiar malady in antiquity. Hippocrates used the term *karcinos* (Greek for *crab*), presumably to convey the pain associated with cancer, which could feel like that caused by the pinching of a crab. Since at least the writings of Hippocrates and later Galen, there has been a tendency to attribute cancer to the influence of the mind over the body. In particular, it was suggested that cancer was caused by a melancholic disposition. The idea that emotional factors and life situations play a role in the development and progression of cancers has continued to be a topic for speculation. One of the earliest studies on the role of life events was conducted in 1893 at the London Cancer Hospital, where it was found that of 250 patients admitted, 62% had experienced a major negative life event before their diagnosis.

In this section, evidence of the impact of psychosocial stress on the occurrence and progression of cancer is discussed. However, by talking about cancer, one ignores the vast heterogeneity that reflects the tissues and cell types that undergo malignant transformation, as well as the potential for growth and metastasis of tumors. One further ignores the stage of the tumor when first discovered, along with characteristics of the individual such as age and general health. It is not surprising that reviews of the literature concerning psychosocial stress and cancer reveal that the number of studies suggesting no effect exceeds the number of studies yielding positive results. As a rule, trends appear for one type of cancer but not for another. Further complicating the picture are reports indicating that although single life events that are stressful appear to be associated with the diagnosis of malignancy, regular exposure to major life stressors seems to reduce the risk of malignancy at least in some samples.

It is also important to keep in mind that psychosocial stress is not the only form of stress experienced by someone diagnosed with cancer. The cancer itself, even before it is diagnosed, is a form of stress and so are treatments such as chemotherapy and surgery. The point here is that the effect of psychosocial stress may be overwhelmed by the burden of the tumor and the trauma of the treatments.

Despite the need for caution, given the preceding considerations, investigators have made a number of interesting observations concerning the psychological and social characteristics of individuals who are more likely to be afflicted with cancer and for whom there is a more rapid progression. Much of what follows is taken from reviews of the literature by Bernard Fox, who has been an astute observer of the evidence.

1. Depression, Psychosis, and Outlook

Depression, as measured using the Minnesota Multiphasic Personality Inventory (MMPI), was found to predict cancer diagnoses 10 years postassessment and to a lesser degree at 20 years postassessment. This evidence is consistent with depression being predictive of cancer diagnosis earlier in life and becoming less of a factor as the sample ages and cancer become more prevalent for a variety of other reasons. However, subsequent studies using different scales to assess depression have not corroborated this finding. Still, some data suggest that patients with both cancer and comorbid depression may progress faster. A complicating factor in more rapid progression is the observation in animal studies that some antidepressant medicines may promote tumor growth.

The possibility that depression reflects early effects of malignancy should not be overlooked. In the 1920s, prospective research found that patients who were later found to have pancreatic cancer exhibited symptoms of major depression first. The figure was 50% for patients with pancreatic cancer but only 17% for those later diagnosed with colon cancer. This sort of evidence may implicate early disturbed function of the pancreas as leading to mental symptoms resembling depression, rather than the reverse.

Psychotic disorders such as schizophrenia, which cause significant stress as indexed by cortisol levels, have been reported to be associated with lower cancer mortality. The possibility that this reflects an effect of pharmacotherapy with neuroleptics has been suggested. However, Fox (1999) has made a case for the fact that increased premature deaths from causes such as suicide, not as prevalent in the general population, artifactually lowers the percentage of deaths from cancer in schizophrenic samples.

Some of the research has suggested that suppression of negative emotions, particularly anger, may be a factor in the probability that cancer will develop and the speed with which it will progress. This has been noted particularly in the case of melanoma. However, this is not completely supported by all the relevant evidence. Fox has suggested that suppression of anger may be more a consequence of cancer diagnosis than a causative factor.

A "fighting spirit" has frequently been regarded as a predictor of better cancer prognosis, whereas a more stoic style has been seen as leading to more rapid progression. For instance, one study found that mortality at 5 years in the group with more "fighting spirit" (i.e., assertive, take charge, optimistic, and goal directed) was lower than that of the stoic, hopeless, and helpless group. Again, this is not entirely consistent with all the relevant evidence.

2. Social Relationships

Social support has been found to be related to cancer incidence in women (lower support equates with higher incidence). In men, the absence of social support appears to be predictive of more rapid progression. Overall, social support seems protective, but this may depend on the type of cancer and stage of disease. According to Fox (1999), the evidence is most consistent with social support being beneficial by retarding cancer progression, but not by preventing its occurrence.

3. Life Events

Regarding life events, the most recent reviews of the research using meta-analytic methodology have found no support for the effect of stressful life events on the incidence of breast cancer in particular. Fox (1999) goes further and concludes, "Stressful events do not occur more often among those who later get cancer, die of it, or survive a shorter time."

4. Psychosocial Treatment and Cancer Survival

Could interventions directed at psychosocial factors after cancer diagnosis be beneficial? The possibility that this could be the case was dramatically strengthened in 1989 when Spiegel *et al.* reported that group psychosocial treatment significantly increased longevity in patients with breast cancer. The optimism was sustained, although the survival of patients receiving treatment did not exceed general population trends; in fact, it was the control group's rather negative course that created the divergence. This was pointed out by Fox (1999), who saw mixed evidence regarding a positive effect of psychosocial treatment on cancer progression. The most recent study in this area, conducted by Goodwin *et al.* (2001) in consultation with Spiegel did not find support for psychosocial treatment as a factor in longevity, although there were improvements in mental well-being for the patients receiving such treatment. Survival in the treatment group of the study by Goodwin *et al.* (2001) resembles that reported by Spiegel *et al.* (1989) in their treatment group. In essence, the difference between the studies reflects a significant difference between the control groups.

5. Immune Mediation of Psychosocial Stress

Direct attempts to investigate aspects of immune function along with psychosocial stressors in patients with cancer are indeed scarce. Levy (1985) observed that women diagnosed with breast cancer who were rated as being better adjusted to the disease (not distraught) and who reported higher fatigue had lower natural killer (NK) cell numbers. In another study, Levy *et al.* (1985) found that those with higher social support tended to have higher NK cell activity. Higher NK cell cytotoxicity measured 1 week after surgery (a time when NK cell activity should decrease from baseline) predicted a longer cancer-free period over the next 5 years.

Fawzy *et al.* (1993) studied patients with melanoma, obtaining mood and immune activity measures, for 6 years after delivering a short-term (10-week) group psychosocial intervention. At 6 years postintervention, mortality in the treatment group was lower. The treatment group had shown a decrease in distress presumably as a result of the intervention and an increase in active coping and in the number of large granular lymphocytes (the NK cell phenotype). The increase in NK cell number was also evident when indexed by the CD56 marker. It was found that those who died earlier had expressed less negative emotion at baseline but were comparable to the more distressed subjects with respect to immune measures. Effects were primarily seen in the male patients. The patients whose NK cell numbers and interferon-α-stimulated NK cell activity increased from baseline over the 6-month period after the intervention remained free of cancer recurrence for a longer interval. However, NK cell activity did not predict survival time. Thus, even though the intervention group experienced the highest survival rate at 6 years (91% vs 71%), it was not necessarily those with the highest NK cell activity or lowest rate of cancer recurrence who were ultimately most likely to survive.

Finally, it is important to emphasize that life events and emotional reactions may have an impact on cancer via multiple pathways. It is not simply a matter of effects on the immune system. Psychosocial stress is known to influence a variety of health-impairing behaviors, particularly use of drugs such as tobacco and food-consumption patterns that may create adverse metabolic conditions. As noted earlier, such variables have powerful effects on the risk of malignancy. Moreover, the neuroendocrine response to psychosocial stress may interfere with DNA repair and increase the likelihood of malignant transformation of cells. Tumor growth is responsive to endocrine stimulation and, thereby, could be influenced by psychosocial stress. Psychosocial stress can affect the individual's willingness to seek medical advice and, thus, alter the possibility that a tumor will be detected early enough to make treatment more effective. Obviously, compliance with treatment can modify its success. Therefore, in essence, although a relationship may be shown to exist between psychosocial stress and malignancy, the involvement of immune mechanisms is not required. Indeed, direct endocrine effects may be more important. For instance, the effect of cortisol is not simply immunosuppression; it appears to enhance the vascularization of tumors and facilitate the use of glucose by malignant cells.

E. Could Psychosocial Factors be Irrelevant?

One might be tempted to dismiss psychosocial factors as irrelevant to cancer. However, extensive animal research demonstrates that stressful circumstances have an impact on cancer incidence and progression. A review of the animal research conducted by Justice (1985) concluded that when tumors were virally induced, they grew faster under stress, whereas non–virally induced tumors appeared to stop growing. Given that the role of infection in human tumors is becoming increasingly evident, the influence of psychosocial stress must remain under consideration. For instance, in the case of cervical cancer, a viral etiology (human papilloma virus [HPV]) is well documented, and so the role of immune defenses may be important. Most Papanicolaou smears (more than 70%) showing dysplasias (a precursor of cancer) give evidence of HPV infection, in contrast to cytologically normal smears (fewer than 4%). Moreover, pharmacologically immunosuppressed individuals (e.g., organ transplant recipients or human immunodeficiency virus–positive individuals) have an increased risk of developing cervical cancer. In contrast, familial or genetic factors appear to have a weak influence at best, whereas lifestyle, by increasing the risk of infection with HPV, plays a significant role in the likelihood that cervical cancer will occur. It is noteworthy that other infectious agents (e.g., *Candida albicans*) are associated with a decreased risk of cervical cancer. Psychosocial stress appears to have an adverse effect in that those with more advanced disease have experienced more negative life events and are more pessimistic, hopeless, anxious, and socially alienated. They have also been found to be more passive, helpless, respectful, and cooperative. Furthermore, alexithymic individuals, who have difficulty expressing emotions, tend to show more evidence of cervical dysplasia, a precursor to cancer. Such subjects tend to have lower $CD8^+$ and $CD4^+$ numbers relative to nonalexithymic individuals without dysplasia, a finding suggestive of immune system involvement. Nonetheless, an important argument against the role of the immune system in cancer protection is the well-established observation that the pharmacological immunosuppression necessary in the case of organ transplant recipients does not augment the posttreatment incidence of all cancers; only certain types of cancers, in particular lymphomas, occur more often.

III. AUTOIMMUNE DISEASES

As the name implies, autoimmune diseases are characterized by some form of specific immune attack directed at healthy tissue with damaging consequences leading to tissue malfunction. Recognition that such disorders existed dates back to the beginning of the twentieth century when the field of immunology was in its infancy. Paul Ehrlich is credited with coining the term *horror autotoxicus* to denote the basic autoimmune process. Despite the early interest, autoimmunity as a possible cause of disease did not really come to the forefront until 1951, when hemolytic anemia was found to be due to an immune attack on red blood cells. In 1957, Hashimoto's thyroiditis was the first organ-specific autoimmune disorder to be recognized. By 1965, autoimmunity was considered an important cause of human disease.

A. CLONAL SELECTION THEORY AND NORMAL AUTOIMMUNITY

Gradual understanding of how the immune system recognized pathogens led to the awareness that there must be a way for immune cells to distinguish foreign antigen from antigen that arises from the body's healthy tissues. The problem of how foreign antigen is recognized led to the clonal selection theory of immune responsiveness. According to this theory, lymphocytes have unique antigen-recognition receptors, which are generated by random rearrangement of specific genes. Therefore, self-reactive lymphocytes must be generated in everyone and must somehow be removed or inactivated to avoid autoimmune reactions. In other words, the individual must become tolerant of antigen arising from healthy tissue. Such tolerance is now understood in terms of both deletion (central tolerance) and inactivation (peripheral tolerance) of self-reactive lymphocytes. Autoimmune reactions occur when tolerance is broken because (1) self-reactive lymphocytes have failed to be deleted or inactivated, (2) self-antigens have some-how become altered so they appear to be foreign, or (3) self-antigens not normally encountered by the immune system become exposed and are treated as foreign. However, it is well established that healthy individuals possess self-reactive T cells and autoantibodies. This has been called *physiological* (or normal) *autoimmunity* and appears essential to the recognition of foreign antigen. Such autoimmunity is not pathological unless it escapes inhibitory regulation. The prevailing opinion is that interference with peripheral tolerance is the most likely cause of autoimmune disorders in individuals with a genetic predisposition.

B. Prevalence of Autoimmune Disorders

Cohen (2000) points out that clonal selection theory, with randomness at its core, would lead to the expectation that autoimmune attack could be directed at any of the tens of thousands of proteins and the hundreds of cell types that characterize the human organism. In theory, there could be hundreds if not thousands of autoimmune disorders. Thus far, the number of recognized disorders is well under 50, suggesting that other factors must be part of the equation, leading to autoimmune disorder. Indeed, fewer than 10 autoimmune disorders account for most of the diagnosed cases.

At the top of the list are disorders affecting the thyroid gland (e.g., Graves' disease, a form of hyperthyroidism, and Hashimoto's thyroiditis, a cause of hypothyroidism), which jointly account for about 50% of all cases of autoimmune disease. Another 20% of the cases involve attack on the joints, predominantly in the form of rheumatoid arthritis (RA). Disorders targeting the kidneys (glomerulonephritis), the liver (autoimmune hepatitis and cirrhosis), the red blood cells (hemolytic anemia), the salivary/lacrimal glands (Sjögren's syndrome), and the skin (scleroderma/uveitis) add another 20% to the cases. Finally, a number of well-known disorders such as myasthenia gravis (affecting neuromuscular transmission), multiple sclerosis (MS) (affecting central nervous system [CNS] transmission), insulin-dependent diabetes (affecting insulin-producing cells), and systemic lupus erythematosus (SLE) (affecting multiple organ systems) collectively account for only about 5% of the estimated 9 million people afflicted with autoimmune disease in the United States.

C. Pathogenic Mechanisms

Self-directed immune attack can be humoral, mediated by autoantibodies, or mediated by T cells responsive to self-antigen. The self-antigen may be expressed in a particular cell type or present in most cell types, so that immune attack will cause an organ-specific or multiple-organ (systemic) disorder.

1. Autoantibodies

The production of autoantibodies can have a variety of consequences. They may disrupt tissue function by causing damage. For instance, in hemolytic anemia, antibodies against red blood cells target them for destruction (hemolysis) by phagocytes or complement. The hyperthyroidism seen in Graves' disease is driven by autoantibodies that act as receptor agonists on the thyroid gland. The motor dysfunction evident in myasthenia gravis reflects the action of autoantibodies that block the acetylcholine receptor at the neuromuscular junction.

2. Self-reactive T Cells

T-lymphocyte involvement in autoimmune disorders occurs even when the ultimate effect is antibody mediated, because often T-cell help is required for antibody production. Nonetheless, some disorders result primarily because of helper T-cell (CD4$^+$) action through the release of cytokines. This appears to be the case in insulin-dependent (type I) diabetes and MS. Primary T-cell involvement is also suspected in inflammatory bowel disorders such as Crohn's disease. However, cytotoxic T cells (CD8$^+$) do not appear to be primary effectors in autoimmune disease.

3. Costimulation

Costimulation via membrane-bound receptors, ligand couplings, and cytokine release is required for immune activation by antigen, including instances in which the activating antigen is derived from healthy self-tissues. Increased IL-4, IL-6, and IL-10 (i.e., a Th2 cytokine profile) can lead to overactivity of B cells that produce autoantibodies. Interferon-γ (IFN-γ) and IL-12 can augment autoimmune attack that is T-cell mediated. In the case of joint inflammation, tumor necrosis factor-α (TNF-α) plays an important role along with IL-8. The latter promotes the accumulation of leukocytes in the joints and upregulates angiogenesis (i.e., blood vessel formation) in the affected joint. Moreover, unregulated healing (i.e., scar tissue formation) can cause additional damage. The overproduction of IL-12 and TNF-α promote cell-mediated (Th1) autoimmune disorders, whereas IL-10 overproduction shifts the balance in the direction of humorally mediated (Th2) diseases. Membrane-bound proteins (e.g., CD80, CD86, CD28, CD40R, and CD40L) that participate in the activation process of lymphocytes are also crucial to the occurrence of an autoimmune response. This has been shown in animal models by blocking the coupling of such molecules with specific antibodies and, thereby, preventing or delaying disease onset.

4. Immune Complexes

Immune complex is the term used to refer to the combination of an antibody with its antigen. Immune complexes must be removed from tissues and kept from accumulating in the circulation and forming deposits throughout the body.

Failure to clear immune complexes can lead to autoimmune disease. Complement fixation to immune complexes facilitates their removal by phagocytes. It also prevents deposits by facilitating their binding to erythrocytes, which transport immune complexes to the spleen and the liver, where resident phagocytes such as macrophages dispose of them. Complement deficiencies, whether genetic, drug induced, or antibody mediated, can cause the buildup of immune complexes and trigger chronic stimulation of immune activity. In turn, this can promote the production of autoantibodies to both cell surface antigens and intracellular antigens of nuclear or cytoplasmic origin. The latter can serve to unleash systemic forms of autoimmunity, which simultaneously affect multiple tissues. The premiere example of multiorgan autoimmunity is SLE, which affects the kidneys, the skin, the joints, the CNS, and other tissues. Consequently, the clinical presentation of SLE varies considerably across patients. SLE is characterized by a multitude of immune abnormalities including elevated autoantibody titers, abnormalities in T-cell function, and complement deficiencies that impair immune complex clearance.

D. Infection Triggers Autoimmunity

As was true in the case of cancer, autoimmune disorders can be triggered by infection. There is also evidence that cancer can cause autoimmune disease. In the case of infection, the mechanism involved has been dubbed *molecular mimicry*, denoting the fact that antigens characteristic of a pathogen, by virtue of being similar in structure to self-antigens, unleash immune responses that may cross-react with healthy self-antigens. It should be noted that molecular mimicry could in principle work to either unleash attack on self-antigens or preclude attack on foreign antigen that resembles self. Molecular mimicry is not the only avenue by which infection can provoke autoimmunity. Infection can cause tissue damage that unmasks cryptic self-antigens (e.g., intracellular proteins) and, thereby, serves to initiate a self-directed immune response. It is also the case that costimulatory signals increased in response to infection, as a collateral phenomenon, may activate anergic self-reactive lymphocytes. Therefore, not every case in which a pathogen is associated with the onset of autoimmunity hinges on molecular mimicry. There is evidence to suggest that a variety of autoimmune disorders seem to be temporally associated with particular infections. These include diabetes (coxsackievirus B infection) and RA (EBV infection). However, the paradigmatic case of autoimmunity triggered by infection is acute rheumatic fever. This condition develops after infection with group A β-hemolytic streptococci. A bacterial protein is the source of an antigen similar to one present in the normal myocardium and can, thus, lead to cross-reactivity. Cross-reactivity in this case has also been observed in the joints, the skin, and the CNS following infection. More is said about the effect of streptococcal infection on neural tissue when the role of immune activity in psychiatric disturbances is discussed in Chapter 10.

A group of disorders known as *spondyloarthropathies*, which includes ankylosing spondylitis (AS), have been associated with enteric gram-negative bacterial infection. It is particularly likely to develop in individuals who possess the specific class I Human Leukocyte Antigen allele B27 (HLA B27). It appears that a microbial antigen closely resembles an antigen derived from HLA-B27 protein, leading to cross-reactivity. Nearly 90% of all cases of AS possess the HLA-B27 allele. AS is not a proven autoimmune disorder, although it does involve inflammation of healthy tissue, particularly involving the joints of the spine. AS can be seriously incapacitating. It is also historically important because it was the disorder that afflicted Norman Cousins, who thought he had successfully reversed the disorder by exposing himself to sustained humorous stimulation. Cousins has detailed his experience in the book *Anatomy of an Illness*.

E. CANCER TRIGGERS AUTOIMMUNITY

As noted earlier in this chapter, another trigger of autoimmune disease is cancer. The cancer-initiated disorders are known as *paraneoplastic syndromes*. They are not due to direct tumor effects or metastases. They result from cross-reactivity of antibodies directed at tumor cells with normal tissues in distal structures. Disorders affecting the eyes, the CNS, and the skin have been observed. Small-cell lung cancer and breast, ovarian, and testicular cancers have been implicated as triggers. Most of the autoantibodies detected in these disorders are directed at cytoplasmic proteins and must somehow gain entry into normal cells to have their deleterious effects.

F. TOXIC CHEMICALS TRIGGER AUTOIMMUNITY

Environmental triggers participate in the process leading to autoimmunity. Infectious agents, as noted earlier, can induce autoimmunity by molecular mimicry or by creating an inflammatory milieu, that is, localized conditions (cytokine levels and expression of costimulatory proteins) that may facilitate activation of self-reactive lymphocytes. Toxins and drugs can alter self-antigens or disrupt the balance of immune-regulatory networks (i.e., the active inhibition of self-reactive lymphocytes) so that immune attack toward self occurs. Toxic chemicals such as mercury and iodine can modify self-proteins enough to make them appear foreign. An important difference between drug-induced and idiopathic autoimmunity is that the former abates when the drug is discontinued.

G. Susceptibility to Autoimmunity

In the final analysis, autoimmune disorders reflect the fact that the immune system must strike a balance between self-tolerance with respect to healthy self and destruction of damaged self and foreign organisms. Individuals are poised in relation to self-tolerance so that some are more susceptible to autoimmunity.

Females are at much higher risk than males. Nearly 80% of all diagnosed cases of autoimmune disease afflict females. In this regard, it is of interest that estrogen enhances the action of glucocorticoids. The corticotropin-releasing hormone (CRH) gene contains an estrogen–sensitive response element. Pregnancy is known to affect autoimmune disorders, an effect that probably reflects the increases in progesterone and estradiol, which affect cytokine expression in the direction that suppresses cell–mediated (decreased TNF-α) and enhances humoral-mediated (increased IL-4 and IL-10) immunity. In addition to pregnancy, the postpartum period, the premenstrual phase, and menopause are all linked to autoimmune modulation, most likely because estrogen and progesterone are at low levels and downregulate the effect of glucocorticoids. Finally, it should be noted that some tissues appear more susceptible, because disorders of the thyroid gland and the joints account for the vast majority of autoimmune disorders.

As is true for all diseases, genetic factors significantly modify risk. In the case of autoimmunity, it has been estimated that the genetic contribution may approach 60%. Animal studies have suggested that as many as 15 distinct genetic loci may contribute to genetic susceptibility. Monozygotic twin concordances have been found to range between 15% and 60%. Specific alleles coding for HLA class II proteins appear particularly influential. They account for more than 50% of the genetic risk. For instance, in the case of insulin-dependent diabetes, nearly 40% of all children afflicted carry HLA-DQ8 and HLA-DQ2 alleles. HLA class II proteins are involved in antigen presentation to T cells (CD4$^+$), and their structure may make it more or less likely that a response to self-antigen could occur. The remaining genetic influence occurs at a variety of levels. For instance, genetic influence either on proteins involved in costimulation, on the complement system, or on target tissue may combine to further facilitate reaction to self-antigen. The importance of target tissue characteristics is suggested by the fact that autoimmune attack may be restricted to a particular organ, although the same self-antigen is present elsewhere. However, the latter also might reflect the effect of a micro-environmental trigger such as a localized infection.

In essence, genetic proclivity may suffice to trigger autoimmunity in some individuals, but in most cases, other factors must be present. Moreover, the progression and the severity of the autoimmune attack appear subject to modula-tion. Gender alters the likelihood of developing an autoimmune disease substantially. Endocrine factors may also be crucial, given that pregnancy influences the severity of autoimmune symptomatology. The possibility of further modulation by psychosocially induced neuroendocrine effects seems reasonable.

H. Psychosocial Stress and Autoimmune Disorder

Autoimmune disorders are clearly heterogeneous conditions and have multifactorial etiologies. The role of psychosocial factors, particularly the influence of life events and personality dispositions, has been frequently investigated in relationship to disease onset and progression. Many of the early studies were retrospective and anecdotal. However, despite the limitations of the literature, psychosocial stress likely plays a role in the onset and course of autoimmune disorders, although the magnitude and direction of the effect may be variable across specific disorders and may depend on other individual characteristics.

Research examining the role of psychosocial stress has been more extensive for some disorders, but not others. A great deal of research has been conducted in RA, which was originally viewed as one of the classic psychosomatic disorders. Much attention is devoted to the research on RA in this section. However, research on autoimmune thyroid disorders is mentioned first because such disorders are by far the most prevalent.

1. Thyroid Disease

Autoimmune thyroid disease has a concordance of 50% in monozygotic twins. Evidence for the role of stress, both physical and psychosocial, is most compelling in Graves' disease, a hyperthyroid condition. As many as 77% of patients with Graves' disease have experienced stressful events before disease onset. One of the early symptoms is increased nervousness, which appears to have been triggered by events such as surgery, pregnancy, traumatic experiences (e.g., automobile accident), or bereavement. An interesting reported observation is the increase of hyperthyroidism in the population of Denmark during the Nazi occupation of that country in World War II. It should also be noted that untreated Graves' disease spontaneously remits in 30–50% of patients.

2. Rheumatoid Arthritis

RA is the second most prevalent autoimmune disorder and has been the subject of much research examining the role of psychosocial factors. In addition to the influence of life events, much attention has been given to the personality characteristics of the afflicted individual.

a. Personality and Relationships

Interest in the personality of the patient is more apparent in early medical writings. More than 1000 years ago, Razi, a Persian physician, concluded that inability to experience or express aggression was at the root of a patient's arthritic condition. Obviously, the situation is not so simple, but research over the last 50 years has tended to support the role of psychological characteristics with respect to RA. Early investigations had noted that patients afflicted with RA were excessively dependent and had difficulty expressing emotions. It was often reported that the precipitating event for disease onset was the loss of or separation from significant others. Arthritics were said to turn hostility inward. Overtly, such patients were seen as calm, composed, and optimistic and rarely expressed or acknowledged feeling angry.

Patients with RA described themselves as perfectionistic, tense, anxious, and introverted compared to their nonafflicted siblings. Patients strove to avoid conflict and, even when reporting bad marriages, still indicated that they did not argue with their spouses. They showed a strong tendency toward self-sacrifice and denial or suppression of hostility. Patients with RA exhibited a much higher rate of depression (46%) than that seen in the general population. Difficulty expressing anger has indeed been noted in many, although there is also a subgroup that is easily angered. Patients who maintain good relationships and have the support of others have better outcomes. This has been supported by recent research indicating that relationship-induced stress exacerbates RA. In about half of the cases of juvenile-onset RA, there was evidence of social disruption (e.g., divorce, separation, death, or adoption) near the time of disease onset. However, the situation is further complicated by observations suggesting that patients who experience anxiety and depression fare better than patients who deny the emotional impact of RA or who externalize their hostility. An intriguing possibility in this regard is that the relative benefit of depression may reside in the lower physical activity and higher cortisol levels seen in some depressed individuals.

b. Genes and Circumstances

Rimon (1969) found that disease onset could be either sudden or slow and gradual. The sudden-onset variety seemed most often associated with psychosocial stress and had no evidence of family history of RA. In the case of the slow gradual-onset variety, a family history of RA was typically present and psychosocial stress was not necessarily implicated. The genetic contribution in RA has been found to be modest. The concordance rate in monozygotic twins is in the vicinity of 15%. Moreover, in monozygotic twins discordant for RA, the affected twin had experienced more psychosocial stress before the onset of illness when compared to the unaffected twin.

RA is more prevalent in populations with lower socioeconomic status, with less education, and a high divorce rate. Usually, the patient tends to attribute disease onset or flare-ups to psychosocial stress, excessive physical activity, or fluctuations in the weather. Flare-ups of RA were documented to have increased after the San Francisco earthquake in the early 1990s. Moreover, it is noteworthy with respect to physical activity that inflammation is reduced in immobilized joints.

c. Immune Mediation

A particularly informative study examined a sample of the unaffected sisters of patients with RA, considering both psychological characteristics and a measure of immune dysregulation known as *rheumatoid factor* (RF). RF is an autoantibody directed at the Fc portion of other antibodies produced by the individual. RF is found in 50–90% of patients with RA, but despite its name, it is not specific and is seen in various other conditions including chronic inflammation. In addition, it becomes increasingly detectable in aging populations. This study, conducted by Rimon, found that the unaffected sisters who were seropositive for RF were better adjusted than the unaffected seronegative sisters. Evidently, psychological disturbance or high psychosocial stress seems necessary to trigger disease even in vulnerable individuals. This appears consistent with the evidence from monozygotic twins. Most recent research has supported the view that higher levels of psychosocial stress are associated with RA in individuals who remain RF seronegative.

Research examining psychosocial stress and immune measures in patients with RA has provided evidence that higher psychosocial stress is associated with a lower helper T-cell/cytotoxic T-cell ratio and a higher percentage of B cells in the peripheral circulation. The percentage of B cells was, in turn, predictive of self-reported exacerbation of RA symptoms.

d. Pregnancy

Pregnancy can be construed as both physical and psychosocial stress. It has been shown to temporarily reduce the symptoms of several autoimmune disorders including RA, which is likely to remit (70–75% of cases) during pregnancy but is also likely to reemerge or originate during the postpartum period. The odds ratio is 5.6 relative to other periods of life and 10.8 for the first pregnancy. These dynamics have been attributed to the fact that during pregnancy, cell-mediated immunity is dampened. Proinflammatory cytokines are suppressed and the balance is shifted in favor of anti–inflammatory cytokines and humoral immunity.

3. Insulin-dependent Diabetes

The trends noted earlier linking psychosocial stress to autoimmune disease are also noted in the case of insulin-dependent diabetes mellitus. In as many as 75% of the cases, the association of psychosocial stress with disease onset appears evident. For instance, one study has documented that in 56% of the cases onset occurred after bereavement or separation, whereas in another 24% of the cases more subtle disruption of social bonds could be detected. As is often the case, this has not been observed in other relevant studies, but animal research clearly documents a significant impact of stress on disease onset in susceptible organisms.

4. Inflammatory Bowel Disease

Early studies tended to implicate difficulty expressing anger and dependency in the case of ulcerative colitis (UC), an inflammatory bowel disease (IBD). UC may begin acutely or gradually, as has been noted for RA. In 16% of the cases, it resolves and never recurs. In most cases (62%), it exhibits a remitting and relapsing course, whereas in 22% of the cases it is persistently progressive and increases the probability of colon cancer. There is familial clustering, and approximately 33% of patients have at least one other relative who is afflicted. Unresolved grief and depression are frequently observed (60%). It is also noteworthy that the prevalence of schizophrenia is significantly increased (6%) in patients afflicted with IBD.

5. Systemic Lupus Erythematosus

The role of psychosocial stress in SLE has not been as extensively investigated. However, evidence suggests that disrupted, unsatisfactory, or conflictual relationships, either of sudden onset or of long duration, tend to precede the emergence of SLE. The loss or even the fear of losing an important relationship has been noted to precede the diagnosis of SLE.

Patients with SLE appear more susceptible to depression than patients with RA. Neuropsychiatric complications are observed in a significant number of patients with SLE. In particular, cognitive impairments and psychotic disturbances are evident in patients with SLE. Electroencephalogram abnormalities, especially over the left temporal areas of the scalp, have been documented. SLE and MS provide compelling evidence that immune dysregulation can impair brain function. Indeed, autoimmunity has been generally invoked as a possible cause of major psychiatric disorders.

SLE, in contrast to RA and other autoimmune disorders, tends to develop or flare up during pregnancy. This appears related to the Th2 predominance during pregnancy, as SLE tends to become attenuated during menopause, when Th2 activation is diminished as a result of declining estrogen and progesterone levels and the downregulation of cortisol action.

6. Multiple Sclerosis

The relationship between disease onset and psychosocial stress has been less clear in MS. Some studies have observed an association comparable to that seen in other autoimmune disorders, but the bulk of the evidence is mixed. However, an association between MS and affective disturbances has been recognized for more than a century and has continued to be evident, particularly regarding depression. Prospective studies have also indicated some relationship between psychosocial stress and relapse. The situation is complicated by some of the evidence indicating that psychosocial stress may reduce disease progression, particularly in individuals who respond with increased glucocorticoid output. In this regard, the increased prevalence of depression in patients with MS may represent a collateral consequence of sustained adrenocortical activity aimed at dampening the disease process.

MS tends to afflict individuals of European ancestry more than others whose recent ancestors came from Asia or Africa. There is an increased incidence in families and the concordance rate for identical twins is approximately 35%. Peak onset is in young adulthood and affects predominantly females. The HLA-DR2 allele is overrepresented in patients with MS. MS is associated with the presence of T cells of the Th1 phenotype in the brain. An increase in blood–brain barrier (BBB) permeability is observed early in the process leading to MS lesions. Modulation of BBB permeability occurs via alterations in the endothelial cells of the brain's microvasculature. Such alterations occur either in response to neural signals that cause the local release of neurotransmitters or peptides or via the action of cytokines and hormones. These signals cause the expression of cell adhesion molecules, which "catch" and guide appropriately activated T cells into the surrounding neural tissue. T cells capable of reacting with neural antigens may be further triggered by encountering antigen presented by microglia or astrocytes. These events ultimately lead to tissue damage and disruption of normal brain function. Finally, a direct correlation has been observed in patients with MS between level of negative affect, particularly anxiety, and the number of helper T lymphocytes. However, this observation, of higher CD4 cell count in more distressed individuals was not related to severity of MS symptoms rated by neurologists.

I. GENDER AND AUTOIMMUNITY

The role of gender in autoimmune disorders deserves further comment. Hormonal factors likely account for this phenomenon. The fact that periodic suppression of cell-mediated immunity is essential to conception raises the possibility that the associated fluxes in cell-mediated immune activity could increase the risk of autoimmune disorders in females. One interesting bit of data concerns RU486, the "morning-after pill" designed to block pregnancy, which happens to antagonize cortisol and progesterone receptors and enhance proinflammatory cytokine production, leading to more pronounced cell-mediated immunity. Nonetheless, it is not entirely possible to dismiss the impact of cultural effects because they constrain how women may express feelings such as anger. Psychosocial factors may also be implicated by observations that suggest that the thoughts and feelings a woman experiences as a result of being pregnant tend to predict whether an exacerbation or a remission of a preexisting autoimmune disorder will occur.

IV. CONCLUDING COMMENTS

Cancer is a form of stress that arises from within. Cancer is fundamentally a disorder that originates at the single cell level. Alterations in genetic expression, as a result of DNA damage, can create conditions in which growth signals predominate and the protective triggering of cell death fails to operate. The genes likely to be involved are numerous and have been dubbed *oncogenes* or *tumor suppressor genes*. Malfunctioning genes can be inherited or result from exposure to DNA-damaging factors such as tobacco smoke, oxygen-free radicals, or infection. However, cancer is more prevalent in technologically advanced societies where mortality from infectious disease is lower. Primary defenses against malignancy appear to be DNA-repair mechanisms or the activation of cell death. The immune system may serve in a backup capacity, but there must be sufficient cell change (i.e., expression of tumor-specific antigen) to break tolerance and provoke an effective immune response.

The role of psychosocial stress in the etiology and progression of cancer has sparked much interest, speculation, and heated debate. The role of negative emotion has long been regarded as an important variable. However, studies have yielded mixed results. Depression has not been consistently linked to cancer risk, although some evidence suggests that it may accelerate progression. There is also some indication for the possibility that cancer, even before it has been diagnosed, may cause depression. Similarly, the suppression of anger, which has been frequently cited as increasing cancer risk, may be more correctly viewed as a consequence of the diagnosis. Possessing a fighting spirit in the face of a cancer diagnosis has yielded mixed results with respect to prognosis. Only a high level of social support has been found to retard cancer progression. Amount of life change has not been consistently found to predict cancer diagnosis, rate of progression, or mortality. However, there is at least a trend for single life-changing events to be associated with cancer diagnosis, whereas chronic conditions of psychosocial stress may even be associated with decreased risk.

It seems fair to ask whether psychosocial stress matters at all in the case of cancer. Animal research suggests that stress matters in virally induced cancers, which grow faster under stress. In contrast, non–virally induced cancers may even stop growing under stress. Thus, when cancers arise by either route, the role of psychosocial stress may not be readily apparent. Finally, the role of psychosocial stress on cancer need not be mediated immunologically, particularly because the immune system is probably not the most important line of defense. Psychosocial stress may have its most significant impact in the genesis and progression of most cancers through effects on endocrine activity and health behaviors, including seeking of medical attention.

Autoimmunity reflects at least in part the failure of self-reactive lymphocytes to be deleted or actively suppressed. Both B lymphocytes and $CD4^+$ T lymphocytes are involved, acting via the production of autoantibodies and the release of cytokines capable of promoting inflammation, tissue injury, and scar formation. An autoimmune process can also be initiated by the failure of immune complexes to be cleared because of complement deficiencies. The risk of specific autoimmune disorders is increased in individuals who carry particular HLA molecules. Toxic chemicals and drugs, which alter self-antigens, can trigger autoimmunity. Autoimmunity can develop as a consequence of infection and cancer.

Females are at higher risk of autoimmune disorders. Moreover, phases of their reproductive function appear to modulate autoimmunity, probably because of hormonal effects on immune function. However, the direction of the effect on symptoms of autoimmunity depends on whether cellular or humoral immunity is the primary culprit in the disorder. For instance, during pregnancy, RA (a cell-mediated disorder) is improved, but SLE (mediated by humoral immunity) tends to be exacerbated.

Psychosocial stress plays a role in autoimmunity, although the direction of the effect may not be simply adverse; a protective role has been documented as well. Psychosocial stress has been shown to increase the risk of autoimmune thyroid disease. Emotional inhibition, especially with respect to the expression of anger, has been associated with RA. Psychosocial stress is seen most often in sudden onset cases, when there is no family history of the disorder. When monozygotic twins are discordant for autoimmune disorders, the afflicted twin has usually experienced more psychosocial stress. Finally, patients afflicted with autoimmune disorders are more susceptible to depression, but it should be noted that those who report anxiety and depression seem to fare better.

What should be apparent by now is that for any disease, there are multiple factors that in different combinations can bring about the same type of abnormal function. The immune system is a key player, because it can cause damage while responding to pathogens, other forms of tissue insult, or more directly in the form of autoimmunity. In addition, psychosocial factors, genetic characteristics, and exposure to nonliving aspects of the environment all play roles. All diseases reflect the contribution of these factors, with temporal arrangements or configurations that can be relatively unique across individuals.

V. SOURCES

Abeloff, M.D., Armitage, J.O., Lichter, A.S., & Niederhuber, J.E. (eds) (2000), *Clinical Oncology*, 2nd ed, New York, Churchill Livingstone.

Andersen, B.L., Kiecolt-Glaser, J.K., & Glaser, R. (1994), "A Biobehavioral Model of Cancer Stress and Disease Course," *American Psychologist*, 49, pp. 389–404.

Cohen, I.R. (2000), *Tending Adam's Garden: Evolving the Cognitive Immune Self*, San Diego, Academic Press.

Cohen, M.J.M., Kunkel, E.S., & Levenson, J.L. (1998), "Association between psychosocial stress and malignancy," in Hubbard, J.R. & Workman, E.A. (eds) *Handbook of Stress Medicine: An Organ System Approach*, Boca Raton, Fla, CRC Press.

Cousins, N. (1979), *Anatomy of an Illness,* New York, W. W. Norton.

Fawzy, F.I., Fawzy, N.W., Hyun, C.S., et al. (1993), "Malignant Melanoma: Effects of an Early Structured Psychiatric Intervention, Coping, and Affective State or Recurrence and Survival Six Years Later," *Archives of General Psychiatry*, 50, pp. 681–689.

Foley, F.W., Miller, A.H., Trangott, V., et al. (1988), "Psychoimmunological Dysregulation in Multiple Sclerosis," *Psychosomatics*, 29, pp. 398–403.

Fox, B.H. (1999), "Psychosocial factors in cancer incidence and prognosis," in Holland, J.C. (ed) *Psycho-Oncology*, New York, Oxford University Press.

Goodin, D.S., Ebers, G.C., Johnson, K.P., et al. (1999), "The Relationship of MS to Physical Trauma and Psychological Stress: Report of the Therapeutics and Technology Assessment Subcommittee of the American Academy of Neurology," *Neurology*, 52, pp. 1737–1745.

Goodwin, P.J., Leszcz, M., Ennis, M., et al. (2001), "The Effect of Group Psychosocial Support on Survival in Metastatic Breast Cancer," *New England Journal of Medicine*, 345, pp. 1719–1726.

Greaves, M. (2002), *Cancer: The Evolutionary Legacy*, Oxford, Oxford University Press.

Hall, S.S. (1997), *A Commotion in the Blood: Life, Death, and the Immune System,* New York, Henry Holt and Co.

Jacobson, D.L., Gange, S.J., Rose, N.R., & Graham, N.M.H. (1997), "Epidemiology and Estimated Population Burden of Selected Autoimmune Diseases in the United States," *Clinical Journal of Immunology and Immunopathology*, 84, pp. 223–243.

Justice, A. (1985). "Review of the Effects of Stress on Cancer in Laboratory Animals: Importance of Time of Stress Application and Type of Tumor," *Psychological Bulletin*, 98, pp. 108–138.

Lakita, R.G., Chiarazzi, N., & Reeves, W.H. (eds) (2000), *Textbook of the Autoimmune Diseases*, Philadelphia, Lippincott Williams & Wilkins.

Levy, S.M. (1985), *Behavior and Cancer,* San Francisco, Jossey-Bass.

Levy, S.M., Herberman, R.B., Maluish, A.M., et al. (1985), "Prognostic Risk Assessment in Primary Breast Cancer by Behavioral and Immunological Parameters," *Health Psychology*, 4, pp. 99–113.

Ligier, S. & Sternberg, E.M. (2001), "The neuroendocrine system and rheumatoid arthritis: Focus on the hypothalamo–pituitary–adrenal axis," in Ader, R. Felten, D.L., & Cohen, N. (eds) *Psychoneuroimmunology*, 3rd ed, San Diego, Academic Press.

Prat, A. & Antel, J.P. (2001), "Neuroendocrine influences on autoimmune disease: Multiple sclerosis," in Ader, R. Felten, D.L., & Cohen, N. (eds) *Psychoneuroimmunology*, 3rd ed, San Diego, Academic Press.

Rimon, R. (1969), "A Psychosomatic Approach to Rheumatoid Arthritis: A Clinical Study of 100 Female Patients." *Acta Rheumatologica Scandinavica,* 13, pp. 1–154.

Rogers, M.P. & Brooks, E.B. (2001), "Psychosocial influences, immune function and the progression of autoimmune disease," in Ader, R. Felten, D.L., & Cohen, N. (eds) *Psychoneuroimmunology,* 3rd ed, San Diego, Academic Press.

Rose, N.R. & Mackay, I.R. (eds) (1998), *The Autoimmune Disease,* 3rd ed, San Diego, Academic Press.

Schiffer, R.B. & Hoffman, S.A. (1991), "Behavioral sequelae of autoimmune disease," in Ader, R. Felten, D.L., & Cohen, N. (eds) *Psychoneuroimmunology,* 2nd ed, San Diego, Academic Press.

Solomon, G.F. (1981), "Emotional and personality factors in the onset and course of autoimmune disease, particularly rheumatoid arthritis," in Ader, R. (ed) *Psychoneuroimmunology,* New York, Academic Press.

Spiegel, D., Bloom, J.R., Kraemer, H.C., & Gottheil, E. (1989), "Effect of Psychosocial Treatment on Survival of Patients with Metastatic Breast Cancer," *Lancet,* 2, pp. 888–891.

Stein, S. & Speigel, D. (2000), "Psychoneuroimmune and endocrine effects on cancer progression," in Goodkin, K. & Visser, A.P. (eds) *Psychoneuroimmunology: Stress, Mental Disorders, and Health,* Washington, DC, American Psychiatric Press.

Turner-Cobbs, J.M., Sephton, S.E., & Spiegel, D. (2001), "Psychosocial effects on immune function and disease progression in cancer: Human studies," in Ader, R., Felten, D.L., & Cohen, N. (eds) *Psychoneuroimmunology,* 3rd ed, vol 2, San Diego, Academic Press.

Van der Pompe, G. (2000), "Neoadjuvant immunostimulation in oncologic surgery," in Goodkin, K. & Visser, A.P. (eds) *Psychoneuroimmunology: Stress, Mental Disorders, and Health,* Washington, DC, American Psychiatric Press.

Visser, A.P., Goodkin, K., Vingerholts, A.J.J.M., et al. (2000), "Cervical cancer: Psychosocial and psychoneuroimmunologic issues," in Goodkin, K. & Visser, A.P. (eds) *Psychoneuroimmunology: Stress, Mental Disorders, and Health,* Washington, DC, American Psychiatric Press.

Weiner, H. (1991), "Social and psychological factors in autoimmune disease," in Ader, R., Felten, D.L., & Cohen, N. (eds) *Psychoneuroimmunology,* 2nd ed, San Diego, Academic Press.

Whitacre, C.C., Cummings, S.O., & Griffin, A.C. (1994), "The effects of stress on autoimmune disease," in Glaser, R. & Kilcolt-Glaser, J.K. (eds) *Handbook of Human Stress and Immunity,* San Diego, Academic Press.

Wilder, R.L. & Elenkov, I.J. (2001), "Ovarian and sympathoadrenal hormones, pregnancy, and autoimmune diseases," in Ader, R., Felten, D.L., & Cohen, N. (eds) *Psychoneuroimmunology,* 3rd ed, San Diego, Academic Press.

Zautra, A.J., Okun, M.A., Robinson, S.E., et al. (1989), "Life Stress and Lymphocyte Alterations among Patients with Rheumatoid Arthritis." *Health Psychology,* 8, pp. 1–14.

CHAPTER 10

Immune Activity and Psychopathology

I. Introduction   180
II. Access to Brain by Pathogens   181
III. Immune Activity within the Brain   181
    A. Lymphocyte Entry   182
    B. Cytokine Effects in Brain   182
    C. Optimal Immune Response in Brain   183
    D. Consequences of Immune Activity
       in the Brain   183
IV. Nervous System Infections
    and Behavior   184
    A. Neurological Disorders   185
    B. Neurodegenerative Disorders   186
    C. Lyme Disease   187
    D. Herpes Viruses   187
       1. Encephalitis   188
       2. Guillain–Barré syndrome   188
    E. Rabies Virus   189
    F. Human Immunodeficiency Virus   189
V. Autoimmunity, Malignancy,
    and Behavior   190
    A. Systemic Lupus Erythematosus   190
    B. Paraneoplastic Disorders   191
VI. Sensing Peripheral Immune Activity   191
    A. Cytokines Alter Mental Processes
       and Behavior   191
    B. Channels of Communication   192
       1. Circulatory Pathways   192
       2. Neural Pathways   192
    C. Functional Significance   193
VII. Sickness Behavior   193
VIII. Behavioral Disorders that Resemble
    Sickness Behavior   194
    A. Chronic Fatigue and Pain   194
    B. Depressive Disorders   195

IX. **Psychiatric Disorders with a Link**
      to Infection   196
      A. Autism and Pervasive
          Developmental Disorders   196
      B. Attention-Deficit/
          Hyperactivity Disorder   197
      C. Childhood Obsessive-Compulsive Disorder
          and Tourette's Syndrome   198
      D. Schizophrenia and Other Psychoses   199
 X. **Concluding Comments   203**
XI. **Sources   204**

I. INTRODUCTION

As mentioned in Chapters 8 and 9, disease states can have more or less specific effects on the function of the central nervous system (CNS) and, thus, induce changes in behavior, which could be transient or enduring. In this chapter, the evidence linking infection and immune activity to changes in cognitive and emotional function is examined in more detail. One would anticipate such effects, along with more profound impairments, in response to infection of the CNS and the resulting activation of immune responses. However, changes in brain function and behavior also occur when immune activity is restricted to the periphery.

Alterations in behavior due to infection or malignancy result from disruption of neural function caused directly by the pathogenic process or indirectly by the activation of immune responses. When pathology does not occur within the CNS, the processes by which it alters behavior are thought to involve mediators such as cytokines or antibodies. Such macromolecules may act locally on peripheral neural tissue and give rise to centrally transmitted neural signals, which affect behavior. Alternatively, cytokines, antibodies, or activated lymphocytes may enter the circulation, penetrate the blood–brain barrier (BBB), and thereby alter brain activity and behavior more directly.

Obviously, insult to other organ systems can result in disruption of brain function and cause behavioral change. However, instances in which immunological or other damage to tissues such as the liver or the lungs is the cause of changes in brain function and behavior are not examined here.

There are many brain disorders in which immune activity has been implicated. Most disorders of this type are regarded as neurological. The involvement of immunological or infectious factors in psychiatric disorders has received some attention but remains a difficult area in which to establish causation in a compelling manner. The distinction between what is regarded as neurological as opposed to psychiatric is largely a matter of tradition. It is partially related to the ease with which neuropathological findings can be observed and to philosophical views regarding mental phenomena. In this chapter, consideration is primarily given to disorders in the realm of psychiatry. However, in the interest of presenting a comprehensive picture, some neurological and neurodegenerative disorders are discussed as well. The presentation begins by examining how pathogens enter the CNS.

II. ACCESS TO BRAIN BY PATHOGENS

Pathogens must enter the body, and as previously mentioned, this occurs across surfaces such as the skin and the mucosal linings of the respiratory, gastro-intestinal, and genitourinary tracts. Some pathogens remain confined to the original site of infection, whereas others have the ability to spread more extensively and, thus, can potentially infect the CNS. Infections of the CNS are relatively rare, even though exposure to pathogens occurs regularly.

Pathogens able to penetrate the CNS are said to be *neuroinvasive*. In the case of viruses and other intracellular pathogens, they must be *neurotrophic* (i.e., capable of infecting cells within the CNS) and *neurovirulent* to actually cause disease.

Most CNS infections are acquired from the blood (hematogenous route). Essentially, the infection grows outside the CNS. In the case of virus infection, this can lead to viremia. Virus particles in the blood may cross the BBB in their free form, often by first infecting the endothelial cells of the brain's vasculature. Virus can also enter the CNS inside infected cells (i.e., Trojan horse mechanism) such as mononuclear leukocytes that are capable of penetrating the BBB. Other infections of the CNS, such as those caused by herpes simplex virus (HSV) or rabies virus, gain access to the CNS by first entering nerve endings in the periphery at the site of the original infection. Such viruses are then able to move intraneuronally and across synapses in a retrograde direction until they gain access to the CNS and move all the way up to higher order neural structures including the limbic system and the neocortex.

III. IMMUNE ACTIVITY WITHIN THE BRAIN

Given that pathogens penetrate into the CNS, immune responses within the CNS must occur. However, the CNS has long been regarded as a privileged site, where immune responses do not occur easily. This has been demonstrated by the fact that foreign tissue transplanted into the CNS is not rejected as handily as in the periphery.

One reason often invoked to explain the protected status of the brain is the existence of the BBB, which acts as an impediment to the entry of leukocytes into the brain. It is also the case that the brain lacks a lymphatic system and that cells within the brain do not normally express the Human Leukocyte Antigen (HLA) molecules necessary for the antigen-specific activation of immune responses. Nonetheless, the potential exists under some circumstances for the activation of immune responses within the CNS. There are cells of bone marrow origin, which populate the CNS. They include perivascular macrophages and microglia. These cells are capable of expressing HLA class II molecules and, thus, can present antigen. Astrocytes and endothelial cells can also be recruited into the role of antigen-presenting cells (APCs). Microglia can act as phagocytes and release a variety of proinflammatory cytokines and trophic factors when activated. Moreover, T lymphocytes activated in the periphery can penetrate the BBB and release cytokines that upregulate HLA expression and cell adhesion molecule expression by the endothelium in the region.

A. LYMPHOCYTE ENTRY

The entry of activated T cells into the CNS does not appear to be antigen specific. However, the retention of such cells within the CNS depends on encountering their specific antigen. Otherwise, the cells recirculate by way of the interstitial fluid draining into the cerebrospinal fluid and then into the cervical lymphatics. This same route can be taken by APCs, which in turn initiate the activation of T cells in the periphery and thereby close the cycle of CNS surveillance by T cells. Activated T cells, which encounter specific antigen within the CNS, promote local proinflammatory activity, as noted earlier in this chapter. They increase the permeability of the BBB so that serum proteins, including cytokines and antibodies, can pass through and be detected at increased concentrations in the cerebrospinal fluid. B lymphocytes are also able to enter the brain at those sites and remain sequestered while producing antibodies.

B. CYTOKINE EFFECTS IN BRAIN

Glial cells can produce and respond to cytokines such as interleukin-1 (IL-1), interferon-γ (IFN-γ), IL-6, tumor necrosis factor-α (TNF-α), TNF-β, transforming growth factor-β (TGF-β), and colony-stimulating factors (CSFs). IL-1 can promote astrocyte proliferation and induce astrocytes to release other cytokines (e.g., TNF-α, IL-6, and CSFs). Astrocytes can also produce complement components in response to IL-1. IFN-γ can upregulate HLA expression when released from activated T cells entering the CNS. TNF-α and TNF-β appear capable of damaging oligodendrocytes, the source of myelin in the brain. CSFs can enhance inflammatory responses in the CNS by activating microglia, the brain's macrophage. Astrocytes are the major source of CSFs in the brain. In contrast, IL-6 appears to suppress glial cell activation. TGF-β operates in an immunosuppressive/ anti-inflammatory fashion and is also released from glial cells. The initiation of these responses seems dependent on infiltration by activated T cells and macrophages into the CNS, which then release cytokines like IFN-γ, thereby activating glial cells. The latter may promote inflammation or curtail it depending on the cytokine profile that becomes predominant. Inflammatory activity within the brain can have negative consequences.

C. OPTIMAL IMMUNE RESPONSE IN BRAIN

Cell-mediated responses must be highly regulated within the CNS because of the potential for destruction of cells that may be virally infected but that will not regenerate. Downregulation of antigen-specific cytotoxic responses is particularly critical when neurons are infected by viruses that are not cytotoxic in and of themselves. Essentially, the immune system must strike a balance between causing neuronal damage and eliminating virus. Nonetheless, effective defense against CNS infection requires the capability to rapidly mount a cell-mediated response, coupled with a robust antibody response aimed at neutralizing extracellular virus and thus preventing further infection. Rapidity is the key to minimize damage and eliminate infection. Minimal reliance on inflammatory responses seems most effective. Antibody in the brain does not normally activate inflammatory responses because of the low concentration of complement and the fact that immunoglobulin G (IgG), an isotype that does not fix complement, is most frequently found in the brain.

D. CONSEQUENCES OF IMMUNE ACTIVITY IN THE BRAIN

Unfortunately, immune activity can destroy not only pathogens but also healthy brain tissue and cause impairment of function. Therefore, pathogens can cause functional disturbances by how the immune system responds to their presence. The response may cause disturbance due to inflammation or due to the production of antibodies. The latter could cause either temporary or permanent impairment depending on whether the antibodies act in a "druglike" capacity (e.g., block or stimulate receptors) or as flags to target cells for destruction. These events can arise because antibodies simply cross-react with healthy neural tissue as a result of molecular similarity between foreign antigen and normal proteins. They can also occur by a process whereby lymphocytes with specificity for normal brain proteins gain access to such antigen, become activated, and produce autoantibodies. Similarly, cell-mediated responses can be directed at self-antigens as a result of the retention of lymphocytes that normally should have been deleted or by molecular similarity between foreign antigen and self-antigen. The result would be inflammatory activity or cytotoxicity, which would impair tissue function either transiently or permanently, when tissue destruction is extensive.

In essence, by virtue of their immune-activating effect, pathogens either in the periphery or within the CNS may cause pathological behavioral changes. Moreover, because immune and brain activities influence each other, changes due to psychosocial factors and infectious agents may spread beyond the system originally affected. Psychosocial events alter brain activity, which then affects immune activity and thereby modifies control over microbial growth. In turn, microbial penetration leads to immune activation, which alters brain activity and modifies psychosocial experience. The issue is no longer one of possible influence, but rather how in a given case the pathological state is initiated and then sustained and how one may best intervene to ameliorate and resolve pathology.

IV. NERVOUS SYSTEM INFECTIONS AND BEHAVIOR

Infections of the nervous system can be caused by viruses, bacteria, fungi, and parasites. As noted earlier in this chapter, such infections are relatively rare because of the multiple barriers that the pathogen must overcome before it can enter the CNS. Infectious agents most often enter the CNS from the blood after having established an infection elsewhere in the body or at sites adjacent to the brain such as the nasal mucosa or the inner ear. Other infectious agents ultimately gain access to the brain by first entering nerve endings in the periphery. CNS infections occur most often in individuals who because of age or health status may lack adequate immune competence. Once infection develops in the CNS, progression to disease can be quite rapid and potentially lethal. However, some pathogens (e.g., viruses, prions, and some bacteria) do not seem to cause effects for long periods after the original infection.

The type of pathogen involved is not easily discerned from the clinical presentation of many CNS infections. Acute encephalitis is most often the result of a viral infection, but other pathogens (e.g., parasites) may be culprits as well. Subacute encephalitis, with its more gradual onset, can result from a wide variety of pathogens. Acute meningitis can be viral, also known as *aseptic meningitis*, which is less severe than bacterial meningitis, a potentially life-threatening infection. The latter is most often associated with infection by *Haemophilus influenzae*, *Neisseria meningitis*, or *Streptococcus pneumoniae*. Other bacterial infections that have clear neuropsychiatric manifestations include those associated with neurosyphilis (general paresis) and Lyme disease.

An important difference between viral infections and those caused by bacteria, fungi, or parasites is the degree to which the infection remains localized.

Viral infections tend to be more widespread, whereas bacterial infections and those caused by other organisms tend to be more localized and can cause brain abscesses. In other words, the infected area may be surrounded by inflammation and become gradually encapsulated as the neural tissue is destroyed and the functions subserved by the brain region are disrupted. Bacterial toxins are also capable of disrupting neural function. This is well documented in the case of botulinum toxin, which blocks the release of acetylcholine in the periphery, thereby disrupting motor behavior. Similarly, tetanus toxin travels into the spinal cord and blocks the release of the inhibitory neurotransmitter glycine, causing uncontrollable muscle spasms. Some toxins act as superantigens capable of causing massive T-cell activation in a relatively nonspecific manner.

Clearly, infection of the CNS can alter behavior to some degree. As noted earlier, this could reflect a direct effect of the pathogen on neural function, even when the pathogen does not kill neurons or accessory cells. Some infections can be rapidly damaging, whereas others may be hardly detectable and persist in latent form. Immune responses may prove effective in eliminating the pathogen, containing damage, or at times may lead to an exacerbation of damage in the process of destroying the pathogen. For instance, chronic infection with *Bornavirus* in susceptible animals can cause a condition primarily marked by behavioral abnormalities, including hyperactivity, overarousal, hyperphagia, and aggression, which eventually lead to lethargy over weeks to months. The initial phase of this behavioral disturbance is associated with inflammation in areas adjacent to blood vessels within the brain. By the time lethargy becomes evident, the inflammation will have subsided. However, the disorder does not develop when infected animals are prevented from mounting an inflammatory response, suggesting that the latter is the primary cause of the behavioral disorder. Often it is not easy to distinguish the relative contribution of pathogenicity and immunogenicity. Again, disruption of neural function can occur when the infection resides outside the CNS proper, for instance, when the infection affects the vascular endothelium of vessels supplying the CNS or even more distal organ systems.

A. Neurological Disorders

Some neurological disorders appear to be postinfectious. They reflect some sort of dysregulation of immune function triggered by the infection. One such condition is Reye's syndrome, which may be triggered by influenza, especially in young individuals who are given aspirin while having the flu. This disorder can be lethal or cause serious disability in the form of mental retardation, severe hyperactivity, seizures, or more focal neurological defects. Another condition of this type is Sydenham's chorea, which can develop following bacterial infection with group A streptococci and is characterized by involuntary movements. Changes in more complex behaviors have also been noted, particularly behaviors that are ritualistic such as those used to define obsessive-compulsive disorders. Still another postinfectious disorder is Guillain–Barré syndrome, which causes paralysis following viral infection.

Infections of the meninges (meningitis) or the brain (encephalitis) are relatively rare complications caused by common pathogens. For instance, more than 100 viruses can cause meningitis or encephalitis. Some of these viruses (e.g., California encephalitis virus) are carried by mosquitoes; such arthropod-borne viruses (arboviruses) can be selective with respect to the neural cells they are able to infect and may not be cytotoxic. Behavioral changes such as irritability and inattentiveness are noted in as many as 15% of patients with evidence of infection.

B. NEURODEGENERATIVE DISORDERS

Viral causes are also suspected in neurological disorders such as Alzheimer's disease, amyotrophic lateral sclerosis, and Parkinson's disease. Viral involvement in Alzheimer's disease may operate by triggering autoimmunity or cross-reactivity between brain proteins and antigen derived from pathogens. Aging is known to increase the amount of detectable autoantibodies. This may be a consequence of thymic involution, which would result in the retention of autoreactive lymphocytes. Brain-reactive antibodies are found in the cerebrospinal fluid of patients with Alzheimer's disease. Whether this reflects a process whereby immune attack on healthy tissue leads to the formation of neurofibrillar tangles and neuritic plaques or whereby the formation of such structures provokes the immune response remains unclear. However, in systemic lupus erythematosus (SLE), there is also an association between brain reactive antibodies and the degree of dementia exhibited by the patient.

In some neurodegenerative disorders, the offending agent is a protein particle known as a *prion*. Prions have been implicated in progressive degenerative disorders of the nervous system. Prions appear to enter the body via the gastrointestinal tract. They do not seem to elicit an immune response. Evidently, the immune system is blind to such an infectious particle. The best-known brain disorders of this type are kuru and Creutzfeldt–Jakob disease. Kuru appears to be contracted by groups who engage in cannibalism of dead relatives as part of their funeral ritual. Creutzfeldt–Jakob disease presents later in life. Its transmission is less well understood. Both disorders may initially manifest as loss of energy, sleeplessness, forgetfulness, confusion, and out-of-character behavior or as a variety of focal disturbances such as aphasia, ataxia, or loss of vision. Encephalopathies of all types present initially with changes in behavior resembling psychiatric disorders, although typically they progress to more flagrant dementias and can cause death. However, some data suggest the possibility that some cases may not progress beyond the initial behavioral changes.

C. LYME DISEASE

Lyme disease is caused by a spirochete, *Borrelia burgdorferi*, an organism similar to *Treponema pallidum*, the spirochete responsible for syphilis. In contrast to syphilis, Lyme disease is vector borne, that is, acquired from tick bites. This disease has been recognized in Europe since the 1920s. Descriptions in the United States have appeared since the 1970s. In 1982, Burgdorfer isolated the causative agent, hence the name.

The spirochete causing Lyme disease is neurotrophic; within a few weeks after peripheral infection, it can reach the CNS and then remain latent for months to years. It exhibits remarkable strain variation. It can avoid immune defenses and even antibiotic treatment by entering fibroblasts, macrophages, or endothelial cells. The spirochete is rarely recovered from the CNS; thus, its effect on neural tissue may be secondary to the triggering of immune activity and inflammatory responses.

Like syphilis, Lyme disease affects multiple organ systems. The disease has dermatological, arthritic, ophthalmological, cardiac, neurological, and psychiatric manifestations. Most studies find that patient's with Lyme disease exhibit irritability, mood lability, or depression (26–66% depending on the study). Children show mood lability and poor school performance. The preponderance of the evidence suggests that Lyme disease tends to be associated with marked mood changes. Other symptoms include fatigue, photophobia, sleep disturbance, spatial disorientation, language-processing deficits, and signs of peripheral neuropathy such as sharp shooting pains and areas of numbness and weakness. The symptoms fluctuate in intensity. They can improve after antibiotic treatment but can just as easily become more problematic at least during the early phase of treatment. In general, case studies have indicated that most forms of psychiatric disturbance may be associated with Lyme disease, for example, depression, mania, panic, anorexia, obsessive-compulsive behavior, and acute schizophrenia-like psychosis. This is similar to what has been observed in neurosyphilis.

D. HERPES VIRUSES

Herpes viruses known to infect humans include HSV types 1 and 2, varicella zoster, Epstein–Barr virus (EBV), cytomegalovirus, and herpes virus types 6, 7, and 8.

After primary infections, the DNA/RNA from herpes viruses can remain latent in the infected cells and become reactivated (i.e., begin producing virus again) at a later time, often in response to stress resulting from other infections, environmental hardships, or emotional reactions to life events. Herpes viruses have been suspected in the etiology of behavioral disorders such as chronic fatigue syndrome (CFS) and schizophrenia.

1. Encephalitis

HSV-1 infection is quite common, with approximately 90% of adults having been infected, but rarely does the HSV-1 infection cause encephalitis, which can be lethal. Infected neurons in the brain tend to be found in the orbitofrontal and temporal lobes, with either a unilateral or an asymmetrical distribution. Such localization may result from viral entry via the olfactory mucosa. When CNS disease develops in adults, it likely reflects a reactivation of the virus rather than primary infection. Initial symptoms may include personality changes, extreme anxiety, hallucinations, or bizarre behavior. Patients then may develop seizures, paralysis, sensory deficits, and eventually coma. However, some individuals may not progress beyond the initial behavioral changes and may return to baseline. Such individuals are usually classified as psychiatrically afflicted and the role of HSV-1 reactivation may not be considered. The general observation that latent viruses are prone to reactivation at stressful times provides a link to the often-noted association between contextual change and psychiatric disturbances. Moreover, life stress by altering immune control over latent viruses may in some individuals lead to reactivation, which depending on the distribution of infected cells within the CNS could cause a variety of mental disturbances.

2. Guillain–Barré Syndrome

Herpes viruses have also been implicated in Guillain–Barré syndrome, an acute inflammatory demyelinating disorder that causes an ascending paralysis and follows a viruslike illness. It has also been noted after vaccinations, surgery, and pregnancy, which are all forms of stress and seem to fit with the notion that a latent infection has been reactivated and caused the paralysis. Despite the dramatic impact on the individual, nearly 85% of patients with Guillain–Barré syndrome show good recovery. Given the inflammatory nature of this disorder, surprisingly corticosteroids are not helpful and may even exacerbate the condition. In contrast, plasmapheresis and intravenous IgG treatments, which are effective in neutralizing or eliminating autoantibodies, appear to be helpful in the treatment of Guillain–Barré syndrome. Thus, the possibility that autoantibodies directed at Schwann cells are involved gains some support.

E. Rabies Virus

Rabies virus is particularly interesting in this context because of its dramatic effect on the victim's behavior. For instance, the infected individual can develop an intense aversion to water (hydrophobia) and exhibit increased irritability and aggressiveness. Typically, rabies is acquired from the bite of an infected animal. However, in rare cases, the infection has been acquired from airborne viruses found in caves densely populated by bats. There have also been reports of infection after corneal transplants from donors, whose deaths were probably caused by rabies. In the case of infection from bites, muscle tissue is initially penetrated by the virus. The acetylcholine receptor may act as the site of binding and viral entry into the cells. Rabies has a relatively variable and occasionally long incubation period (from 15 days to as long as 4–6 years), which has been attributed to the persistence of infection in muscles, ultimately leading to viral entry into nerve endings and retrograde transport to the CNS. Rabies develops in only approximately 15% of infected individuals who have not undergone treatment.

The onset of clinical disease in individuals infected with rabies virus begins with malaise, anorexia, headache, and fever. The most severe course initially causes nervousness. In a matter of hours to days, confusion, autonomic arousal, and bizarre behavior emerge. Periods of intense agitation and aggressiveness may be interspersed with episodes of drowsiness. Also, a paralytic clinical course may be precipitated by a different strain of the rabies virus that is less easily transmitted. The paralytic form causes a progressive flaccid paralysis, which resembles Guillain–Barré syndrome.

Despite the dramatic changes in mental function induced by rabies virus and the frequent mortality, neuropathological findings can be elusive, suggesting that disruption of function must occur by interference with subcellular events without causing neurotoxicity. The fact that there are documented instances of recovery attests to the reversibility of the pathological process. Rabies appears as one of the better documented instances in which disruption of neuronal function by the virus itself is the principal reason for the behavioral change.

F. Human Immunodeficiency Virus

Human immunodeficiency virus (HIV) has been extensively studied. It is found in the brain of 90% of those afflicted with acquired immunodeficiency syndrome (AIDS), even when no neurological symptoms are evident. The virus may enter the brain in free form during the viremia of early infection or it may travel into the brain in infected mononuclear cells (e.g., monocytes). Once in the brain, it tends to infect cells of hematopoietic origin such as microglia but has also been found in astrocytes, ependymal cells, and neurons. Disruption of neural function may reflect the effects of viral proteins (e.g., glycoprotein 120 [gp120]), increased release of proinflammatory cytokines from glial cells, reduced production of neurotrophic factors, or cross-reactivity between antibodies to HIV and neural proteins. More generally, a disruption of the BBB can lead the entry of neurotoxins from the general circulation. Remarkably, despite significant disruptions in brain function by HIV, neuropathology not attributable to other opportunistic infections is not as pronounced as in the case of other viral infections of the CNS.

V. AUTOIMMUNITY, MALIGNANCY, AND BEHAVIOR

As noted in Chapter 9, autoimmune disorders can produce transient disturbances in behavior that appear as psychiatric disorders. This section examines the occurrence of such disturbances in the case of SLE.

A. SYSTEMIC LUPUS ERYTHEMATOSUS

Mental disorders in conjunction with SLE were first reported in the literature as early as the late 1800s. In 1895, Osler observed that psychiatric symptoms frequently accompany SLE. The percentages of patients affected by mental disturbances have ranged from 3% to 82%, with a median of 41%. What is particularly difficult to discern in a given case is whether the symptoms have their origin in the autoimmune process affecting the brain directly or affecting other organ systems. Further complicating the situation are the patient's reaction to knowing that he or she has a chronic disease and the effects of pharmacological treatments.

The mental disturbances seen in SLE may result from inflammation of small blood vessels in brain (vasculitis), deposition of immune complexes in choroid plexus or in the brain, or the action of specific antibrain antibodies. In practice it is very difficult to be sure that systemic or other organ effects of the disease are not the cause of mental changes. It simply cannot be done based on behavior alone. In fact, it may require a level of monitoring that is typically not necessary to clinically manage the disorder. One study that attempted to separate cases of mental disturbance with accompanying metabolic disturbances or signs of other organ involvement from those without found that approximately 50% of the psychiatric symptoms occurred without evidence of other organ involvement. Of these, an unknown number may be attributable to the effect on the individual of the knowledge of being afflicted with a complex disease that is chronic, multifaceted, and unstable. Again, it should be noted that some drugs used to treat SLE, such as antihypertensives and corticosteroids, are known to produce psychiatric symptoms.

B. PARANEOPLASTIC DISORDERS

It is also important to highlight in this context that malignancy can affect mental function without occurring intracranially or as a result of metastasis or other organ involvement. These disorders are called *paraneoplastic neurologic disorders*, although psychiatric symptoms (e.g., depression, paranoia, and hallucinations) are often observed (30–50%). Disorders of this type have been recognized since the 1930s, but only within the last 20 years has evidence become available that antibodies to the tumor are capable of cross-reacting with healthy neural tissue, thereby causing damage. Cancers that produce such disturbances include small cell lung carcinoma and breast, ovarian, and testicular cancers. The psychiatric symptoms or neurological abnormalities (e.g., seizures) often predate recognition that cancer has developed. In the case of testicular cancer, some patients may be relatively young and are likely to be seen as suffering from drug-induced or functional psychoses.

VI. SENSING PERIPHERAL IMMUNE ACTIVITY

Activation of immune responses in the periphery has been found to alter brain electrical activity and neurotransmitter levels within specific structures. Such observations along with well-documented behavioral changes in response to infection have underscored the possibility that the immune system acts in a sensory capacity to alert the organism of the presence of pathogens.

A. CYTOKINES ALTER MENTAL PROCESSES AND BEHAVIOR

There is striking evidence that cytokines administered in large doses as part of cancer treatment can cause disturbances of mental function that resemble psychiatric disorders. Cytokines such as IFN-α and IL-2 have been found, depending on dosage, to cause major mental status changes in as many as two-thirds of patients receiving the treatments. The changes observed include delirium (a state of confusion), disorientation, irritability, delusions, agitation, fatigue, anorexia, malaise, and depression. Many of these changes are also seen in response to infection, thereby supporting the role of immune activity in behavioral disturbance and psychopathological conditions. A related, often overlooked finding is that psychiatric medicines have effects on the immune system. For instance, lithium (used to treat bipolar disorder) is a potent stimulator of granulopoiesis. In addition, neuroleptic drugs (used to treat schizophrenia) have been found to induce the production of autoantibodies.

In this context, it is also of interest that there is some evidence, which has remained controversial, that certain foods can cause significant behavioral disorders. Some researchers have referred to such a phenomenon as "food allergy," but it is typically not the case that allergic responses can be elicited by skin tests. However, the argument has been made that the effects are initiated in the gastrointestinal tract as the food molecules interact with gut-associated lymphoid tissue in susceptible individuals. How this activity ultimately is transmitted to the brain remains unclear. There are obviously neural and circulatory routes. An additional possibility involves disruption of gastrointestinal function, leading to the entry of molecular species not normally found in the circulation and with the potential to have neurotoxic effects. The latter possibility is regarded as neural dysfunction secondary to other organ malfunction and is not considered within the purview of this discussion. Only neural changes directly caused by immune activity are considered relevant in this context.

B. CHANNELS OF COMMUNICATION

1. Circulatory Pathways

How brain activity is ultimately altered by large molecules released during immune activation has been the subject of some debate, particularly because the BBB is expected to block entry of such macromolecules into the areas in which neurons reside. Available data suggest that the effect of cytokines on brain activity and behavior may begin with interactions between cytokines and glial cells in areas of the brain where the BBB is less evident, such as the circumventricular structures (e.g., organum vasculosum of the lamina terminalis). The glial response, in turn, would affect nearby neurons and thus initiate the change in brain activity. For instance, there is evidence that IL-1 can induce the release of prostaglandin E_2 from astrocytes, which then alters the activity of thermoregulatory neurons in the hypothalamus and induces fever.

Other mechanisms whereby cytokines may alter brain activity have received support as well. Cytokines may bind to endothelium and trigger a cascade of changes that could affect brain activity in the immediate area. Active transport of cytokines across the endothelium has also been reported. Another pathway begins in the periphery, with cytokines altering neural activity at the site, which is then transmitted to the brain. Actually, the latter seems more plausible under normal circumstances because the concentration of cytokines will be higher at the site of immune activation and would probably be more effective in initiating activity as opposed to when cytokines are diluted in the circulation.

2. Neural Pathways

Maier and Watkins (1998) have provided an excellent discussion of how communication between the periphery and the CNS may occur in the case of

IL-1. They view the vagus nerve, which broadly innervates visceral organs, including lymphoid tissues, as an essential pathway for transmission into the CNS. It has been shown that behavioral changes induced by infection or IL-1 injected peripherally can be blocked by severing the vagus nerves. It is further evident that even if the vagus terminals do not always express IL-1 receptors, there are structures in their vicinity (referred to as *paraganglia*) that do have significant capacity to bind IL-1. It is believed that neurotransmitters contained within the paraganglia are released onto the vagal terminals by IL-1 and initiate a neural response that travels into the brainstem. IL-1 is also found in the brain and it seems to be selectively released when peripheral IL-1 activates the vagal pathway. Thus, IL-1 participates in the modulation of brain activity by initiating a signal in the periphery that then triggers IL-1 release centrally from neurons and results in the behavioral changes that have been associated with infection.

C. FUNCTIONAL SIGNIFICANCE

Maier and Watkins (1998) have also noted that pathogens were the first predators that threatened early forms of life. They argue that as organisms have become more complex, the evolution of the response to threat at any level has retained components of the original response to pathogens. In other words, threatening circumstances, whether in the form of attack by a predator or something more symbolic in nature, will still activate neural networks that modulate immune responses. Such mobilization of the immune system may be seen as preparatory in cases in which injury and infection may be a possibility. However, in instances in which such outcomes are improbable (e.g., undergoing examinations), the response constitutes an irrelevant mobilization or even a potentially maladaptive consequence of how the body has evolved to respond to perceived threat. Thus, there is overlap in adaptive systems so a response triggered by pathogens is also at least partially evoked when facing other threats even if the response is not particularly adaptive in the context at hand.

VII. SICKNESS BEHAVIOR

Activation of immune responses due to the presence of pathogens evokes the release of cytokines, such as IL-1, TNF-α, and IL-6, which initiate what is known as the *acute-phase response* of inflammation aimed at fighting infection. The *acute-phase response* has metabolic and behavioral consequences. Sick animals including humans become feverish, lose their appetite, appear lethargic or sleepy, lose interest in grooming behaviors, and seem to experience pain more easily. These changes appear to facilitate immune responsiveness but also inhibit the ability of some pathogens to multiply. It seems probable that such behavior serves to curtail the spread of infection as well.

Animals that develop fever tend to survive infection better. Fever therapy has been used in earlier periods to fight infections. For instance, syphilis was once treated by infecting the patient with malaria. Older individuals who have difficulty producing fever have a higher mortality in response to infection. Loss of appetite seems at first glance detrimental, but the evidence that force feeding animals suffering from bacterial infections results in higher mortality argues otherwise. Sleepiness, low activity, and absence of grooming are all behavioral changes that seem capable of conserving energy and augmenting temperature. Sickness behavior occurs in response to infection consistently across species and irrespective of the lethality of the infection. Finally, components of sickness behavior, such as fever, are also triggered by noninfectious stressors such as painful stimulation or social isolation. The relative nonspecificity of sickness behavior is reminiscent of Selye's general adaptation syndrome in response to stress.

VIII. BEHAVIORAL DISORDERS THAT RESEMBLE SICKNESS BEHAVIOR

The foregoing discussion has made it clear that immune activity has an impact on brain activity and modifies behavior in the process of fighting infection. Sickness behavior is the result, and interestingly, aspects of sickness behavior resemble some forms of behavioral disturbance. Such behavioral disturbances can be conceptualized as reflecting the failure of an adaptive response to turn off once its beneficial consequences have ceased to be necessary. Alternatively, the response may remain active because the infectious agent has continued to pose a threat, and thus the organism has become stuck in a sickness mode, even if the infection is unlikely to cause damage.

The similarity of sickness behavior to various maladaptive conditions is striking. Lethargy and malaise characterize disorders such as CFS, fibromyalgia (FM), and depression, raising the possibility that such disorders occur when aspects of sickness behavior fail to turn off.

A. Chronic Fatigue and Pain

Viral infections may be followed by prolonged periods of lassitude. Over time such a phenomenon has gone by a variety of labels. In 1869, the term *neurasthenia* was used to denote "a lack of nerve force" as the basis of the fatigue. More recently, the term *CFS* has come into vogue. The condition is diagnosed after at least 6 months of increased fatigue, in association with mild fevers, myalgias, sleep disturbance, and depression. The condition was thought to be due to EBV infection. Recent research has not supported such a link but has demonstrated a variety of immunological alterations in patients with CFS. The suggestion has been made that such patients have a diminished ability to control latent infection with herpes viruses. They have a diminished lymphocyte-proliferation response to mitogens and decreased natural killer (NK) cell activity.

FM is another condition that resembles sickness behavior. It is primarily characterized by widespread pain of musculoskeletal structures. The condition has been linked to chronic infection and to autoimmune disease. It is often observed in patients afflicted with SLE, rheumatoid arthritis (RA), and Sjögren's syndrome. It is predictive of the development of other rheumatoid disorders. The pain is associated with a feeling of swelling in the afflicted areas. A number of other symptoms occur with FM, including severe fatigue, poor sleep, cold intolerance, and low-grade anxiety and depression. The symptoms can flare up in response to physical exertion, exposure to cold, lack of sleep, and psychological challenges. The latter are all forms of stress. Steroid treatment is usually not beneficial.

B. DEPRESSIVE DISORDERS

Aspects of depression, such as anorexia and general lack of interest, resemble features of sickness behavior and again raise the possibility of at least a partial failure to terminate the sickness response. In fact, there is some evidence that after episodes of viral infection, individuals describe themselves as more depressed. Also consistent with this line of reasoning is that administration of cytokines (e.g., IFN-α) to humans can induce depression. Moreover, depressed individuals have been found to have higher levels of acute-phase proteins, proinflammatory cytokines (e.g., IL-6), and soluble IL-2 receptor, as well as higher white blood cell counts, as is often seen in the early response to infection. Essentially, depression appears to be associated with activation of some aspects of nonspecific immunity. However, antidepressant treatment does not necessarily normalize cytokine profiles even when effective at ameliorating the depression.

The situation is more complicated because there is also evidence that depressed patients are immunologically less reactive than controls. They have lower proliferative responses to mitogens and lower NK cell activity. In addition, although depressed patients have some of the features characteristic of sickness behavior including hypersomnia (e.g., in atypical depression), most depressed patients have difficulty sleeping. Insomnia affects the immune system; particularly, cytokine production is affected and NK cell activity is decreased as a result of sleep deprivation. In addition, depressed individuals may be more heavy users of alcohol and tobacco. These are substances that independently have been shown to affect measures of immune function. Thus, until more carefully conducted studies sort out the influence of such confounding variables, the notion that immune changes in depression resemble those evident during sickness behavior must remain tentative.

IX. PSYCHIATRIC DISORDERS WITH A LINK TO INFECTION

Being linked to infection does not imply that a specific pathogen is the cause of some particular psychiatric disorder or that all instances of a given psychiatric disorder have a clear link to infectious or immunological factors. There is enough research and clinical observation to make such possibilities entirely unlikely. In contrast, there appears to be more than enough evidence to make it entirely likely that a proportion of cases presenting with a particular psychiatric disorder will have infectious or immunological factors underlying the pathophysiology leading to the behavioral manifestations that define the disorder.

A. AUTISM AND PERVASIVE DEVELOPMENTAL DISORDERS

Autism was first described by Kanner in 1943 as an affliction of very young children characterized by a seeming disregard for other humans, impaired language development, a proclivity for becoming engrossed in repetitive movements or activities, and a strong aversion to change or novelty. Such children and others with similar pervasive developmental disorders are also often mentally retarded.

A well-documented finding implicating infection in the etiology of autism is the increased prevalence of the disorder (100 times higher than in the general population) in children born to mothers who were infected with the rubella virus during gestation. Prenatal herpesvirus infections have also been observed to increase the risk that an autism-like syndrome will develop. These observations are consistent with recent evidence demonstrating that prenatal brain injury is associated with evidence of intrauterine T-cell activation and increased levels of proinflammatory cytokines.

In the case of children who have been developing normally well into the second year of life and suddenly begin to exhibit autistic-like behavior, a role for infection has been suggested as well. One speculative scenario implicates bacterial infection of the gastrointestinal tract and the possibility that bacterial toxins may be taken up by vagal nerve endings and gain access to the brain where they prove to be neurotoxic. Alternatively, the infection may cause changes in the permeability of the gastrointestinal tract mucosa and allow potentially neurotoxic peptides to enter the circulation. However, the latter would constitute an instance of mental disturbance secondary to other organ dysfunction and not necessarily an effect of immune activity directly on the brain.

A controversial finding, which has been widely publicized despite extensive negative epidemiological evidence, is the association between the onset of autistic behaviors and inoculation of toddlers with the measles, mumps, and rubella (MMR) vaccine. Needless to say, this has been a source of great concern for many parents. The finding originally reported in 1998 by Wakefield *et al.* indicated that after vaccination some children were found to have behavioral changes associated with significant inflammation of the gastrointestinal tract mucosa. It was speculated that the autistic disorder developed as a result of the absorption of molecular entities not normally allowed to enter the circulation. Other possibilities include neurally mediated effects on the brain in response to massive inflammation of the gut or the release into the circulation of large quantities of cytokines such as IL-2. The latter is consistent with the observation that autistic children have higher serum levels of IL-2 in comparison to normal or mentally retarded children. It is also noteworthy in this context that food allergies have been thought to cause some cases of autism as well, again implicating some disruption of the gastrointestinal tract mucosa.

In addition, there have been some retrospective reports by parents that the onset of autistic behaviors followed treatment of infection with multiple rounds of broad-spectrum antibiotics. Such treatment can disrupt the normal flora of the intestines and potentially allow antibiotic-resistant strains of bacteria to flourish and provoke a more intense inflammatory response. Alternatively, it is also known that as bacteria die in response to antibiotics, they become fragmented and may temporarily create greater proinflammatory stimulation, thereby increasing the potential for damage to healthy tissue. In essence, there are a number of plausible scenarios whereby an infectious agent, viral or bacterial, could initiate a sequence of events leading to the alterations in brain function that underlie the behavioral manifestations of autism-spectrum disorders.

B. ATTENTION-DEFICIT/HYPERACTIVITY DISORDER

Attention-deficit/hyperactivity disorder (ADHD) is characterized by over-activity, distractibility, impulsivity, and difficulty adhering to external demands or guidelines for behavior. The disorder has been recognized for a long time. However, a noticeable increase in the number of children who exhibited this type of behavior was reported after the encephalitis epidemic of 1917/1918. The disorder came to be known as "postencephalitic behavior." Despite this clear link between infection and a behavioral disturbance closely resembling ADHD, there has been a tendency to ignore the role of infection in ADHD and to put more emphasis on genetic or toxic (e.g., lead poisoning) causes. Indeed the possibility that genetic susceptibility may reflect response to infection is generally overlooked.

Research on mice has shown that neonatal infection with HSV-1 causes overactivity and inability to inhibit behaviors even when they cease to be adaptive. These deficits tend to be chronic and persist into adulthood. Histological examination of the mice's brains provides evidence of inflammation in the striatum and the cerebellum. These structures have also been implicated in ADHD by neuroimaging studies of children.

As has been noted for autism, the onset of symptoms of ADHD has been linked retrospectively to recurrent infections during early childhood and the need for repeated courses of antibiotic treatment. "Food allergy" has received some support as well, particularly in younger children. Clearly, various observations raise the possibility that infection or immune activation may bring on at least some cases of ADHD.

C. CHILDHOOD OBSESSIVE-COMPULSIVE DISORDER AND TOURETTE'S SYNDROME

Childhood-onset obsessive-compulsive disorder is characterized by recurrent thoughts, impulses, or images that are experienced as distressing and that lead to repetitive behaviors (e.g., hand washing) or mental acts (e.g., repeating a word silently) to reduce the mounting distress. Tourette's syndrome is a tic disorder, characterized by sudden, rapid, recurrent, nonrhythmic movements or vocalizations (e.g., grunts) that occur in bouts throughout most days.

These disorders have been linked to infection with group A β-hemolytic streptococcal bacteria. This organism has been previously shown to cause rheumatic fever, a condition that manifests with arthritis and inflammation of the heart valves and myocardium. It can lead to congestive heart failure or gradually cause heart valve damage. There can also be neurological involvement in the form of Sydenham's chorea, which appears as purposeless nonrhythmic movements, muscle weakness, and emotional disturbances. Chorea in response to streptococcal infection can be delayed for weeks to months. The chorea may be the only manifestation of rheumatic fever. Initially only emotional and behavioral symptoms may be evident, with choreiform movements appearing later. For instance, a case reported by Swedo (1994) illustrates the progression of symptoms. The patient was a 9-year-old girl who experienced an abrupt onset of nightmares and anxiety about being separated from her mother. These symptoms appeared several weeks after a relatively minor upper respiratory tract infection and about 1 month before the appearance of choreiform movements. She began to wash her hands excessively. She would not touch the family's dog or use public restrooms. She needed frequent reassurance that she had not been exposed to rabies and that she was not a bad person. Her affect was labile, laughing happily at one moment but quickly shifting to crying with much sadness or screaming in an enraged state. At school, she was further described as hyperactive, irritable, distractible, and inattentive. Approximately 2 weeks after the onset of the emotional disturbance, she seemed clumsier and exhibited jerky movements. This was followed by deterioration in her handwriting, difficulty speaking and doing daily activities such as brushing her teeth. She also made repetitive facial grimaces and exhibited ticlike movements of her shoulders and neck. However, the progression observed in this case is not seen in all cases, and it is now believed that a subset of patients only appear to be afflicted by behavioral disturbances and do not exhibit the arthritis, carditis, or full-blown choreiform disorder.

As noted earlier, the evidence suggests that antibodies produced in response to streptococcal infection cross-react with normal proteins in the brain, particularly within the basal ganglia, and thereby give rise to the behavioral disturbances. This is consistent with other research implicating the basal ganglia in disorders such as ADHD, obsessive-compulsive disorder, and various disturbances of motor function including tics, as in Tourette's syndrome. Clearly, not everyone afflicted with streptococcal infection goes on to develop rheumatic fever or the associated behavioral manifestations. A high percentage of individuals who succumb to these disorders appear to carry a marker, designated *D8/17*, on a high proportion of B lymphocytes in comparison to healthy individuals (87% vs 17%, respectively). These findings serve to further underscore the role of infection in the behavioral disorders of children and have given rise to the term *pediatric autoimmune neuropsychiatric disorders associated with streptococcal infection* (PANDAS). There may, however, be other pathogens capable of similar effects on behavior.

D. Schizophrenia and Other Psychoses

The modern understanding of schizophrenia and related psychoses begins with the work of Emil Kraepelin and Eugen Bleuler in the late 1800s to the early 1900s. Kraepelin used the term *dementia praecox* to denote the early onset of a dementing process characterized by delusions, hallucinations, and idiosyncratic behaviors. Bleuler advanced the view that underlying such behavioral manifestations was a fundamental fragmentation of mental processes. Thus, Bleuler coined the term *schizophrenia*, which means "split mind."

Schizophrenia is estimated to afflict about 1% of the population across cultures. The major symptoms that characterize schizophrenia are delusions, hallucinations, disorganization of thought processes, inappropriate affect, and behavior or mannerisms that are odd. For instance, schizophrenics report unrealistic beliefs, about which they are entirely certain and not able to reconsider. They experience voices speaking to them when no one is present. Their conversation can be difficult to comprehend because the logic is unconventional, words may seem randomly assembled, or associations are highly idiosyncratic. They can exhibit emotional responses without clear precipitants (e.g., laugh for no apparent reason). In some patients, their movements or lack of movements (catatonia) are clearly odd. Some patients also exhibit a general social withdrawal (negative symptoms) marked by apathy, flat affect, and diminished communication.

The diagnosis of schizophrenia, according to the American Psychiatric Association's *Diagnostic and Statistical Manual of Mental Disorders*, 4th edition (DSM-IV), requires that some of the aforementioned symptoms must begin in early adulthood or earlier. The symptoms must be sustained for at least 6 months and there should be clear evidence of functional impairment in everyday life. It is also stipulated that the symptoms must not be due to substance toxicity or a recognizable medical condition. The latter is important for this discussion because by definition schizophrenia should not be diagnosed when there is clear evidence of infection, malignancy, or autoimmunity. Nonetheless, there are instances of subclinical infection or immune deregulation that may solely manifest with the symptoms of schizophrenia and would, therefore, be appropriately labeled as such rather than as psychosis due to a general medical condition.

The role of infection in psychosis was suggested in early writings noting that "insanity" seemed to wax and wane in the population from year to year without obvious moral causes. Both Kraepelin and Bleuler acknowledged that infectious agents could cause at least some cases of schizophrenia. Karl Menninger (1928, p. 479) reviewed the available evidence concerning infection and reached the following conclusion:

> Infectious disease ... in certain persons breaks the integrative fabric of conscious-
> ness and releases a psychological regression of various degrees and types ... but appar-
> ently most frequently the delirious and the schizophrenic.... The particular type of
> psychotic picture revealed in a particular case ... probably depends on the kind of mental
> substructure preexisting and not demonstrably on the kind of toxin (or infection).

Much evidence has accumulated since the observations of early pioneers in this field to suggest that infection, most likely acquired prenatally, may play a role in the causation of schizophrenia. The evidence is largely circumstantial. Large-scale epidemiological studies based on detailed population records have indicated that offspring born to mothers who were pregnant during influenza epidemics in their region have a higher prevalence of schizophrenia. However, studies that documented influenza in the mother, as opposed to the community, have not found an increased prevalence of schizophrenia in the offspring. One possibility for this discrepancy could be that in the case of influenza subclinical infection of the mother may be more detrimental to the fetus.

Other evidence suggesting prenatal infection includes the consistent observation that the prevalence of schizophrenia is influenced by season of birth; that is,

more cases are diagnosed in individuals born during winter months (February to April), particularly at northern latitudes and in urban areas. This evidence has been interpreted as supporting the role of a transmissible agent such as a virus due to the higher population density and the tendency to remain indoors during late fall and winter months. More recent evidence has shown elevated cytokine levels during pregnancy in the mothers of offspring who later develop psychotic disorders.

The case for some type of prenatal insult is bolstered by neuroimaging findings in schizophrenics, which reveal subtle deviations in brain structure such as decreased hippocampal volume and enlarged lateral ventricles. This has been interpreted as evidence of interference with the process of neural development. However, as previously noted, fetal infections can cause severe disruptions of brain development that render the individual disabled from early life (e.g., autism). Among the viruses that can be quite devastating are rubella virus and cytomegalovirus, a major cause of mental retardation. HSV-2 and HIV are also capable of infecting the neonate during the birth process and can have a significant impact on brain function.

A question that naturally arises is why is it that schizophrenia usually emerges during young adulthood in seemingly healthy individuals when it is thought that prenatal events are determinant? The ideas that have been proposed include the possibility that neurodevelopmental events during adolescence (e.g., net loss of synapses in the frontal cortex) may unleash symptoms by unmasking the inability of prenatally damaged structures (e.g., hippocampus) to keep pace with the flow of information. Another speculative idea considers that the changes in brain organization during adolescence may trigger the activation of latent viruses or retroviral genomes within specific brain structures that either directly or by evoking immune responses disrupt brain function and bring on the symptoms of schizophrenia.

The search for direct evidence of specific viral infection in schizophrenia and other major psychoses has been ongoing since the 1950s. Most studies have focused on specific antibodies to herpesviruses in the serum and cerebrospinal fluid. The results have been mixed, although most studies yield negative results. Those that detect differences between schizophrenic and control groups usually find increased serum and cerebrospinal fluid antibody titers in the schizophrenics and other patient groups. This same trend has been observed at least in the serum of normal individuals under stress. Antibodies to *Bornavirus* have been studied as well and were found to be elevated most often in patients with affective disorders such as depression and bipolar disorder.

The search for viral antigens in the brain of schizophrenics postmortem has not yielded encouraging results. Likewise, efforts to detect viral genomes using polymerase chain reaction techniques have not been successful. However, these are very complex searches and negative results are not definitive proof that viral involvement is not a factor. Even when infection is no longer detectable, it could have still triggered pathology. One way this could occur is through an autoimmune process. In fact, the roles of genetic vulnerability, onset in early adulthood, and course marked by remission and relapse hint at parallels between schizophrenia and autoimmune diseases. Such a possibility began receiving attention in the 1930s and was further advanced by Heath in the 1950s. Essentially antibrain antibodies could have resulted in response to an original infection and then become periodically activated and cause clinical disease, as seems to happen in autoimmune disorders. Although this remains a viable hypothesis, studies searching for antibrain antibodies have tended to yield equivocal results. Moreover, the fact that neuro-leptics used to treat schizophrenia increase the production of autoantibodies raises the possibility that the finding of antibrain antibodies is due to treatment. Finally, an interesting observation is the negative association between RA and schizophre-nia. In essence individuals with RA are at low risk for schizophrenia, which has led to the speculation that the bias toward RA may involve genetic loci that are involved in risk for schizophrenia as well.

Studies of immunological abnormalities independent of infection are among the earliest with respect to schizophrenia. Early studies observed that schizophrenics had an increased number of lymphocytes. Some of the lymphocytes had an irregular appearance, as was later noted in the case of infectious mononucleosis. Schizophren-ics were also found to have a blunted response to antigen injected intradermally; that is, they had a diminished delayed-type hypersensitivity response. Other studies have reported preliminary associations to specific HLA class II molecules. A more con-sistent finding is the increased number of $CD4^+$ T cells in schizophrenics, particu-larly the T-cell subtype that enhances antibody production ($CD4^+$ $CD29^+$). In fact, the $CD4^+$ known as *T-suppressor inducers* ($CD4^+$ $CD45^+$) appears reduced. This is similar to what has been observed in autoimmune disorders. In essence, schizo-phrenics exhibit a Th1/Th2 imbalance in favor of Th2 activation. Also noteworthy is that the increased number of T cells is related to chronicity of schizophrenia and that an increased ratio of $CD4^+$ to $CD8^+$ cells is more often found in individuals with a positive family history of schizophrenia. The presence of negative symptoms has been associated with higher levels of IgG in the cerebrospinal fluid and the response to treatment tends to be better if the T-cell count is lower.

The observation that administration of cytokines such as IL-2 can produce significant mental disturbance resembling psychoses has led to speculation that IL-2 overproduction within the gastrointestinal lymphoid tissue may be a factor in schizophrenia. Such overproduction may be the reason that peripheral blood lymphocytes of schizophrenics have diminished production of IL-2 in response to mitogens. Irrespective of the merit of such speculation, the blunted IL-2 response is still frequently observed. Furthermore, it is associated with higher levels of autoantibodies and soluble IL-2 receptor, the latter being a reliable indicator of immune activation found in transplant rejection, acute infection, and autoimmunity. However, these immune abnormalities are not specific to schizophrenia.

Finally, the experts have remained cautious in interpreting the available evidence. Solomon in 1981 underscored the possibility that the brain condition known as *schizophrenia* may be the cause of the immune abnormalities. Yolken and Torrey (1995) have pointed out the complexity in this area of research. Infection may operate *in utero* or in early postnatal life and then be undetectable. One or several common infectious agents may need to interact with genetic vulnerability in a way that is also affected by the timing and the precise cerebral location of the infection, or the infection may trigger an autoimmune process. In conclusion, they note, "There are no studies that provide a definite link between an infectious agent and these diseases [schizophrenia and bipolar disorder]. In this respect, [they] are similar to multiple sclerosis or Parkinson's disease, both of which are suspected of having a viral etiology, but definite proof is still lacking" (p. 140).

X. CONCLUDING COMMENTS

It is generally well established that infection by any pathogen can occur without evidence of clinical disease. In some cases very severe clinical manifestation will occur, although in most cases only some symptoms of disease may be evident. In effect individual variability is the norm in the realm of infectious disease. Therefore, it seems entirely reasonable and consistent with much evidence that infections capable of affecting the brain will give rise to diverse symptom profiles, with some cases being characterized exclusively by psychiatric disturbances.

The symptoms may be due to the action of the pathogen directly on the brain or may reflect the effect of immune responses directed at the pathogen. Symptoms may also result from pathogen-triggered immune dysregulation, which may take the form of chronic inflammation or enhanced production of antibodies directed at healthy brain tissue. The specific clinical symptoms observed probably reflect the ease with which particular brain functions can be disrupted. For instance, attention is highly vulnerable to even minor disruptions of brain activity. More pronounced disruptions can further interfere with certain integrative aspects of information flow in the brain crucial to the modulation of emotional reactions or the maintenance of a realistic outlook. Even more extensive disruption of brain activity will cause significant neurological disturbances such as seizures and coma.

In essence, infection and immune activity can be a source of disruption to brain function, which in combination with genetic characteristics, exposure to toxic substances, other organ dysfunction, and psychosocial events may give rise to psychiatric disorders. Indeed, the role of infection in psychiatric disorders has probably been underestimated.

XI. SOURCES

Allen, A.J., Leonard, H.L., & Swedo, S.E. (1995), "Case Study: A New Infection-Triggered, Auto-immune Subtype of Pediatric OCD and Tourette's Syndrome," *Journal of the American Academy of Child and Adolescent Psychiatry*, 34, pp. 307–311.

Andreasen, N.C. (1999), "Understanding the Causes of Schizophrenia," *New England Journal of Medicine*, 340, pp. 645–647.

Anisman, H. & Merali, Z. (2002), "Cytokines, Stress, and Depressive Illness," *Brain, Behavior, and Immunity*, 16, pp. 513–524.

Baumann, H. & Gauldie, J. (1994), "The Acute Phase Response," *Immunology Today*, 15, pp. 74–80.

Bennett, R.M. (2000), "Fibromyalgia," in Lahita, R.G. (ed) *Textbook of Autoimmune Diseases*, Philadelphia, Lippincott Williams & Wilkins.

Buka, S.L., Tsuang, M.T., Fuller-Torrey, E., et al. (2001), "Maternal Cytokine Levels during Pregnancy and Adult Psychosis," *Brain, Behavior, and Immunity*, 15, pp. 411–420.

Camara, E.G. & Chelune, G.J. (1987), "Paraneoplastic Limbic Encephalopathy," *Brain Behavior and Immunity*, 1, pp. 349–355.

Capuron, L., Hauser, P., Hinze-Selch, D., et al. (2002), "Treatment of Cytokine Induced Depression," *Brain, Behavior, and Immunity*, 16, pp. 575–580.

Chen, R.T. & DeStefano, F. (1998), "Vaccine Adverse Events: Causal or Coincidental?" *Lancet*, 351, pp. 611–612.

Crnic, L.S. (1991), "Behavioral consequences of virus infection," in. Ader, R., Felten, D.L., & Cohen, N. (eds) *Psychoneuroimmunology*, 2nd ed, San Diego, Academic Press.

Crow, M.K. (2000), "Superantigens," in. Lahita, R.G. (ed) *Textbook of Autoimmune Diseases*, Philadelphia, Lippincott Williams & Wilkins.

Dantzer, R. (2001), "Cytokine Induced Sickness Behavior: Where Do We Stand?" *Brain, Behavior, and Immunity*, 15, pp. 7–24.

Darnell, R.B. (1999), "The Importance of Defining the Paraneoplastic Neurologic Disorders," *New England Journal of Medicine*, 340, pp. 1831–1832.

De Beaurepaire, R. (2002), "Questions Raised by the Cytokine Hypothesis of Depression," *Brain, Behavior, and Immunity*, 16, pp. 610–617.

Denicoff, K.D., Rubinow, D.R., Papa, M.Z., et al. (1987), "The Neuropsychiatric Effects of Treatment with Interleukin-2 and Lymphokine Activated Killer Cells," *Annals of Internal Medicine*, 1017, pp. 293–300.

Duggan, P.J., Maalouf, E.F., Watts, T.L., et al. (2001), "Intrauterine T-Cell Activation and Increased Proinflammatory Cytokine Concentrations in Preterm Infants with Cerebral Lesions," *Lancet*, 358, pp. 1699–1700.

Fallon, B.A. & Nields, J.A. (1994), "Lyme Disease: A Neuropsychiatric Illness," *American Journal of Psychiatry*, 151, pp. 1571–1583.

Fuller-Torrey, E. & Yolken, R.H. (2001), "The Schizophrenia–Rheumatoid Arthritis Connection: Infectious, Immune, or Both?" *Brain, Behavior, and Immunity*, 15, pp. 401–410.

Ganguli, R., Brar, J.S., & Rabin, B.S. (1994), "Immune Abnormalities in Schizophrenia: Evidence for the Autoimmune Hypothesis," *Harvard Review of Psychiatry*, 2, pp. 70–83.

Giedd, J.N., Rapoport, J.L., Leonard, H.L., et al. (1996), "Case Study: Acute Basal Ganglia Enlargement and Obsessive-Compulsive Symptoms in an Adolescent Boy," *Journal of the American Academy of Child and Adolescent Psychiatry*, 35, pp. 913–915.

Goodkin, K., Shapshak, P., Fujimura, R.K., et al. (2000), "Immune function brain and HIV-1 infection," in Goodkin, K. & Visser, A.P. (eds) *Psychoneuroimmunology Stress, Mental Disorders, and Health*, Washington, DC, American Psychiatric Press.

Goodnick, P.J. & Klimas, N.G. (eds) (1993), *Chronic Fatigue and Related Immune Deficiency Syndromes*, Washington, DC, American Psychiatric Press.

Hart, B.L. (1988), "Biological Basis of the Behavior of Sick Animals," *Neuroscience and Biobehavioral Reviews*, 12, pp. 123–137.

Hart, D.J., Heath, R.G., Sautter, F.J., Jr, et al. (1998), "Antiretroviral Antibodies: Implication for Schizophrenia, Schizophrenia Spectrum Disorders, and Bipolar Disorder," *Biological Psychiatry*, 45, pp. 704–714.

Hinze-Selch, D. & Pollmächer, T. (2001), "In Vitro Cytokine Secretion in Individuals with Schizophrenia: Results, Confounding Factors, and Implications for Future Research," *Brain, Behavior, and Immunity*, 15, pp. 282–318.

Irwin, M. (2001), "Depression and immunity," in Ader, R., Felten, D.L., & Cohen, N. (eds) *Psychoneuroimmunology*, 3rd ed, San Diego, Academic Press.

Irwin, M. (2002), "Effects of Sleep and Sleep Loss on Immunity and Cytokines," *Brain, Behavior, and Immunity*, 16, pp. 503–512.

Johnson, R.T. (1998), *Viral Infections of the Nervous System*, 2nd ed, Philadelphia, Lippincott–Raven Press.

Kent, S., Bluthé, R.-M., Kelley, K.W., & Dantzer, R. (1992), "Sickness Behavior as a New Target for Drug Development," *Trends in Pharmacological Sciences*, 13, pp. 24–28.

Klimas, N.G., Salvato, F.R., Morgan, R., & Fletcher, M.A. (1990), "Immunologic Abnormalities in Chronic Fatigue Syndrome," *Journal of Clinical Microbiology*, 28, pp. 1403–1410.

Machon, R.A., Mednick, S.A., & Huttunen, M.O. (1997), "Adult Major Affective Disorder after Prenatal Exposure to an Influenza Epidemic," *Archives of General Psychiatry*, 54, pp. 322–328.

Maes, M., Van der Planken, M., Stevens, W.J., et al. (1992), "Leukocytosis, Monocytosis, and Neutrophilia: Hallmarks of Severe Depression," *Journal of Psychiatry Research*, 26, pp. 125–134.

Maier, S.F. & Watkins, L.R. (1998), "Cytokines for Psychologists: Implications for Bidirectional Immune-to-Brain Communication for Understanding Behavior, Mood, and Cognition," *Psychological Review*, 105, pp. 83–107.

Meijer, A., Zakay-Rones, Z., & Morag, A. (1988), "Post-Influenzal Psychiatric Disorder in Adolescents," *Acta Psychiatrica Scandanavica*, 78, pp. 176–181.

Menninger, K.A. (1928), "The Schizophrenic Syndrome as a Product of Acute Infectious Disease," *Journal of Neurology and Psychiatry*, 20, pp. 464–481.

Merrill, J.E. & Murphy, S.P. (1997), "Inflammatory Events at the Blood Brain Barrier: Regulation of Adhesion Molecules, Cytokines, and Chemokines by Reactive Nitrogen and Oxygen Species," *Brain, Behavior and Immunity*, 11, pp. 245–263.

Mortensen, P.B., Pedersen, C.B., Westergaard, T., et al. (1999), "Effects of Family History and Place and Season of Birth on the Risk of Schizophrenia," *New England Journal of Medicine*, 340, pp. 603–608.

Muller, N. & Ackenheil, M. (1995), "The immune system and schizophrenia," in Leonard, B. & Miller, K. (eds) *Stress, the Immune System and Psychiatry*, New York, John Wiley & Sons.

Narayan, O., Herzog, S., Frese, K., & Rott, R. (1983), "Behavioral Disease in Rats Caused by Immunopathological Responses to Persistent Borna Virus in the Brain," *Science*, 220, pp. 1401-1403.

Perry, S. & Müller, F. (1992), "Psychiatric aspects of systemic lupus erythematosus," in Lahita, R.G. (ed) *Systemic Lupus Erythematosus*, 2nd ed, New York, Churchill Livingstone.

Rapaport, M.H. & Miller, N. (2001), "Immunological states associated with schizophrenia," in Ader, R., Felten, D.L., & Cohen, N. (eds) (2001), *Psychoneuroimmunology*, 3rd ed, San Diego, Academic Press.

Reed, W.P., Johnson, D.R., & Davis, L.E. (1998), "Nervous system infection," in Brillman, J.C. & Quenzer, R.W. (eds) *Infectious Disease in Emergency Medicine*, 2nd ed, Philadelphia, Lippincott–Raven Press.

Rothermundt, M., Arolt, V., & Bayer, T.A. (2001), "Review of Immunological and Immunopathological Findings in Schizophrenic," *Brain, Behavior, and Immunity*, 15, pp. 319–339.

Schwarz, M.J, Chiang, S., Müller, N., & Ackenheil, M. (2001), "T-Helper-1 and T-Helper-2 Responses in Psychiatric Disorders," *Brain, Behavior, and Immunity*, 15, pp. 340–370.

Singh, V.K., Warren, R.P., Odell, J.D., & Coles, P. (1991), "Changes in Soluble Interleukin-2, Interleukin-2 Receptor, T8 Antigen, and Interleukin-1 in the Serum of Autistic Kids," *Clinical Immunology and Immunopathology*, 61, pp. 448–455.

Smith, R.S. (1991), "Is Schizophrenia Caused by Excessive Production of Interleukin-2 and Interleukin-2 Receptors by Gastrointestinal Lymphocytes?" *Medical Hypotheses*, 34, pp. 225–229.

Smith, R.S. (1991), "The Immune System Is a Key Factor in the Etiology of Psychosocial Disease," *Medical Hypotheses*, 34, pp. 49–57.

Solomon, G.F. (1981), "Immunologic abnormalities in mental illness," in Ader, R. (ed) *Psychoneuroimmunology*, New York, Academic Press.

Swedo, S.E. (1994), "Sydenham's Chorea: A Model for Childhood Autoimmune Neuropsychiatric Disorders," *Journal of the American Medical Association*, 272, pp. 1788–1791.

Swedo, S.E., Leonard, H.L., Mittleman, B.B., et al. (1997), "Identification of Children with Pediatric Autoimmune Neuropsychiatric Disorders Associated with Streptococcal Infections by a Marker Associated with Rheumatic Fever," *American Journal of Psychiatry*, 154, pp. 110–112.

Tuomanen, E. (February 1993), "Breaching the Blood Brain Barrier," *Scientific American*, pp. 80–84.

Uthgenannt, D., Schoolmann, D., Pietrowsky, R., et al. (1995), "Effects of Sleep on the Production of Cytokines in Humans," *Psychosomatic Medicine*, 57, pp. 97–104.

Volz, R., Gultekin, S.H., Rosenfeld, M.R., et al. (1999), "A Serologic Marker of Paraneoplastic Limbic and Brain Stem Encephalitis in Patients with Testicular Cancer," *New England Journal of Medicine*, 340, pp. 1788–1795.

Wakefield, A.J., Murch, S.H., Anthony, A., et al. (1998), "Ileal-Lymphoid-Nodular Hyperplasia, Nonspecific Colitis, and Pervasive Developmental Disorder in children," *Lancet*, 351, pp. 637–641.

Yolken, R.H. & Torrey, E.F. (1995), "Viruses, Schizophrenia and Bipolar Disorder," *Clinical Microbiological Reviews*, 8, pp. 131–145.

CHAPTER 11

Immune Function Enhancement

I. **Introduction**   **207**
II. **Beliefs, Suggestion,**
   and Expectations   **208**
   A. Hypnosis   209
   B. Placebo Effect   210
   C. Nocebo Effect   211
   D. Expectations Are Patterns
     of Neural Activity   211
   E. Expectation, Conditioning,
     and Immune Activity   212
III. **Social Engagement**   **213**
IV. **Expression of Emotion**   **213**
V. **Sleep and Relaxation**   **214**
VI. **Exercise and Physical Activity**   **215**
   A. Leukocytes   215
   B. Acute-Phase Proteins, Antibodies,
     and Cytokines   216
VII. **Nutrition**   **217**
   A. Malnutrition   218
   B. Individual Factors
     and Nonlinear Effects   218
   C. Antioxidants: Vitamins and Minerals   219
   D. Lipids   220
   E. Modulation by Hormones   220
VIII. **Concluding Comments**   **221**
IX. **Sources**   **222**

I. INTRODUCTION

Earlier chapters in this book have described how the immune system operates upon encountering pathogens and how its activity is subject to the influence of signals originating within other bodily systems, including the brain.

A guiding assumption has been that such intercommunication has evolved due to its adaptive value, although there are circumstances in which health can be impaired by virtue of the intercommunication between the brain and the immune system. Specifically, psychosocial stress can impair immune function and facilitate disease, whereas immunological activity can affect the brain, alter psychosocial function, and may even cause mental disorders. Essentially, one is always dealing with trade-offs, but as long as the balance is in favor of benefit at the population level, any adaptation is likely to endure.

Much of the previous discussion has emphasized the evidence that psychosocial factors may modify immune function in ways that impair health. However, this chapter focuses on evidence that a variety of personal beliefs and practices, including social activity, expression of emotion, relaxation habits, physical exercise, and nutritional choices, have beneficial effects on immune function. It is assumed that such immune function enhancement is protective of health, although the possibility of adverse consequences exists as well. The latter is rooted in the crucial importance of balance when it comes to immune activity. *Less* is not synonymous with *bad outcomes* and *more* is not always the way to good health.

This exploration of immune-enhancing activities first addresses how beliefs and suggestion combine to create states of expectation. What an individual expects has effects on bodily systems, which must be considered when gauging the impact of any other intervention.

II. BELIEFS, SUGGESTION, AND EXPECTATIONS

Beliefs are complex mental phenomena that serve as guiding principles in the process of human life. They provide ready explanations about how nature works and establish who are credible sources of information. Beliefs are acquired through exposure to the predominant culture, and its religious traditions. Beliefs can also reflect idiosyncratic experiences or result from the workings of imagination.

An individual's belief system opens the door to suggestion from credible sources (e.g., clergy, doctor, and guru) regarding what to *expect* from an intervention or remedy. More generally, expectations are constantly formulated as a result of ongoing experience and without the need for explicit suggestion. Expectations can operate without conscious awareness. As this presentation proceeds to discuss interventions that modify immune function, keep in mind that beliefs and suggestion shape expectations, which, in turn, are players in the organisms' physical response to any intervention.

The power of suggestion has been most dramatically demonstrated under hypnosis, a state in which one individual is highly susceptible to suggestion given by the hypnotist. However, suggestion operates without the need for the hypnotic state. One can simply modify expectations by straightforward statements or even through the process of conditioning, in which case expectation arises as a result of previous association between events.

The expectation that one will experience relief or symptom reduction can create the impression that a neutral intervention has a beneficial effect. This is commonly known as the *placebo effect*. There is also a negative placebo, or nocebo, effect, which refers to the situations in which expectation of detrimental effects or disease serves to promote such an outcome.

A. HYPNOSIS

Hypnosis is fundamentally a process rooted in the willingness of one individual to participate with another in a relationship marked by the belief in the power of suggestion delivered by a qualified source. The hypnotist is thereby able to shape the experience of the subject to an extent that overrides the constraints of reality. There is no more dramatic demonstration of the power of the mind bolstered by the power of relationship. However, not all individuals are equally susceptible to influence under hypnosis. This is not simply a matter of not believing in the power of hypnosis; it essentially shows that like anything else suggestion has a variable and limited influence.

There is abundant evidence that suggestion under hypnosis can alter basic aspects of experience such as the sensation of pain. Analgesia induced via hypnosis has been repeatedly demonstrated. Suggestion under hypnosis has also been found to influence bodily responses mediated by the immune system. This includes modification of various medical conditions in which inflammation is prominent. Hypnosis can alleviate urticaria, psoriasis, atopic dermatitis, and asthma. It has also been found to heal warts caused by localized infection.

Hypnotic suggestion has been found to attenuate immediate hypersensitivity reactions, which as the name implies occur rapidly in response to an offending agent (e.g., allergen) or a mediator of inflammation such as histamine. Suggestion is effective irrespective of whether it is given under hypnosis or not. Furthermore, the hypnotic state can diminish the flare-and-wheal response without the need for specific suggestion. The mechanism underlying these effects appears to be a diminution of the neurogenic component of inflammation, which modifies regional vascular permeability at the site of stimulation. The latter is suggested by histological findings from biopsy material showing that hypnosis does not affect infiltration of the site by immune cells, although inflammation is reduced.

Suggestion given under hypnosis has also been found to modulate delayed hypersensitivity reactions, which are mediated by T cells through the release of proinflammatory cytokines in response to specific antigen that has been previously encountered by the individual. This reaction develops many hours after encountering antigen because of the need for extensive processing of the antigen. Moreover, the effect of suggestion under hypnosis appears to be dependent on the type of antigen used to induce the reaction. For instance, reactions to tuberculin are more readily modified than those of varicella zoster—again, an indication that the impact of suggestion is variable. In addition, the impact of hypnotic suggestion can be more pronounced if the conditions at the time of sensitization (original exposure) are recreated at the time of testing (reexposure). A word of caution is indicated regarding the effect on delayed hypersensitivity because the more carefully conducted studies have yielded negative results.

Finally, the influence of hypnotic suggestion has been examined with respect to more global measures of immune function such as leukocyte counts in the peripheral circulation and the response of lymphocytes to stimulation with mitogens *in vitro*. The available evidence indicates that the hypnotic state tends to decrease the leukocyte count and enhance the responsiveness of lymphocytes to mitogen stimulation. The most reliable findings concern an increase in secretory immunoglobulin A (sIgA) levels and neutrophil adherence. All of these effects may be a reflection of the relaxed state induced by hypnosis rather than the result of suggestion per se. In this context, it is worth noting that the effects of hypnotic suggestion are generally more pronounced in individuals suffering from clinical disorders, who may be more motivated to find a source of relief.

B. PLACEBO EFFECT

The history of attempts to heal the sick until relatively recent times has been overwhelmingly a history of what has come to be known as the placebo effect. In other words, the belief by both the afflicted person and the healer that some potion or ritual has the power to cure could bring about such an outcome. This is certainly a likely factor in the case of many treatments provided under the rubric of alternative medicine even to this day.

The term *placebo* first appears in the medical literature in 1811 to refer to a remedy given more to please than to treat the patient. The word is Latin, meaning "I shall please." The placebo effect has been found to benefit between 20% and 60% of participants in a wide range of research studies, although under some circumstances the effect may be lower or higher than this range. The occurrence of the placebo effect stands as clear evidence of the unity of mind and body. Unfortunately, this has not always been seen in a positive light. Some biomedical researchers, who prefer a disembodied conceptualization of mental processes, find it problematic. In addition, those who would like effective treatments to be entirely exogenous and independent of belief find the placebo effect to be a nuisance. The latter group is often concerned with the potential for financial gain from patented treatments. Given the power of the placebo effect, any treatment or intervention must be shown to exceed its impact before it can be regarded as specifically beneficial. Moreover, such a demonstration must be based on double-blind random-assignment methodology.

C. Nocebo Effect

The nocebo effect refers to the observation that expectation of sickness or even death may promote such an outcome—in effect that the power of belief operates also in negative outcomes. Because the nocebo, or negative placebo, effect is often overlooked, certain communications about health-endangering conditions or practices may serve to promote disease. For instance, well-intentioned communications about the impact of diet or psychosocial stress may create expectations that could lead to disease. In essence, the belief that a poor diet or feelings of loneliness can cause disease may help bring about disease at least in some individuals.

An extreme example of the nocebo effect occurs when the belief that one is about to die serves to bring on death. Voodoo death, though questionable in many instances, has been regarded as a powerful example of the influence of belief over bodily processes that sustain life. Obviously, the individual must be absolutely certain that a curse or hex can have the power to bring on death. Thus, culture and one's immersion in it prove pivotal to the occurrence of physiological changes incompatible with life. However, more generally any communication that promotes expectations of health or disease can influence the occurrence of health or disease.

D. Expectations are Patterns of Neural Activity

Expectations denote the anticipation of experience. They reflect the history of the individual's experience encompassing teachings received from culturally credible sources and the regularities noted by the individual through exposure to specific patterns of stimulation. All these influences, which include conditioning, translate into a biasing of momentary readiness to process events and generate specific responses. Most fundamentally, expectations are patterns of activity within the brain that are not simply contained within higher order neural structures but in fact flow into progressively peripheral structures so they may initiate changes in how the organism senses and responds to input at all levels. Recognition that an expectation is a spatiotemporal pattern of neural activity serves to demystify how it can lead to changes in bodily function. Essentially, there is extensive neuroanatomical evidence demonstrating that neural connectivity between any two structures is bidirectional; therefore, changes in a central pattern of activity can be transmitted to the periphery and alter what transpires there. The change in peripheral activity will, in turn, be sensed centrally and may lead to a self-sustaining pattern of activity.

E. EXPECTATION, CONDITIONING, AND IMMUNE ACTIVITY

Given that expectation appears capable of having bodily effects, is there any evidence that such effects are detectable independent of the individual's self-report? The answer is simply yes. For instance, studies have shown that when subjects expect to receive a stimulant drug but are given placebo, they manifest changes in heart rate and blood pressure as if they had received the stimulant. Changes in immune activity have been documented as well. Placebos have been found to induce or alleviate inflammatory reactions of the skin.

An interesting demonstration of the impact of expectation can be found in a study in which subjects were led to believe that treatment using ultrasound stimulation would relieve pain after wisdom teeth extraction. Pain relief was equally good irrespective of whether the ultrasound machine was turned on or off, provided that both the patient and the doctor believed that the machine was operating. Most importantly, the effect was noted not only in the amount of reported pain, but also in terms of decreased inflammation.

There have also been cases of cancer regression attributed to what the patient expects will happen. An often-cited case was reported by Klopfer in 1957. The patient was a man suffering from lymphoid cancer, who had developed massive tumors. Despite the advanced stage of his cancer, he was allowed to participate in a trial of the drug krebiozen, which had received a great deal of media attention as a miracle cure for cancer. The patient's response to the drug was considered miraculous; the tumor masses regressed to the point that they were no longer detectable and he regained a healthy appearance. Unfortunately, when media coverage of the drug became negative, the patient's condition deteriorated and massive tumors reemerged. The doctors, operating on the assumption that the patient's expectation had been responsible for the original response, attempted to manipulate the patient into the expectation that a higher dose of the drug would work. In fact, he was actually injected with sterile water. Again, dramatic improvement was noted, which lasted until reports appeared in the press indicating that according to the American Medical Association, krebiozen was worthless against cancer. At this point, the patient became sick once more, his tumors grew massive, and he died shortly thereafter.

Actually, the most compelling evidence that expectation can modify immune responsiveness comes from the classical conditioning studies conducted by Ader and others. Conditioning has been regarded as a methodology for inducing expectations, which probably underlie the placebo effect under some circumstances. Thus, the repeated finding that immune responsiveness can be modified by classical conditioning further demonstrates that expectation has the power to alter immune function. In essence, it appears possible to suppress or enhance immune responsiveness by presentation of a neutral stimulus (e.g., a placebo) that has been previously associated with a stimulus that naturally alters immune responsiveness (e.g., a drug or an antigen).

III. SOCIAL ENGAGEMENT

It is important to underscore that the effects of hypnosis and those attributed to placebo are fundamentally the result of communications from others. They require the participation of other humans, whom the subject views as credible, and thus reflect the power of social engagement.

The importance of social engagement or more generally social interaction has been abundantly documented by epidemiological studies that demonstrate that those who have more social contact are more likely to retain health and live longer. These beneficial effects have been assumed to result from social support, although under some conditions and for some individuals social interaction may have adverse effects.

Investigation of the effect of social engagement or social support on immune function has been limited. Social support has often been subsumed under the rubric of stress management along with other interventions such as health education, pain management, and cognitive modulation of distressing experience.

Studies in which some form of social support was undertaken yield mixed results. Across studies, the only consistent finding is an increase in the lymphocyte count. This is noteworthy because at least some retrospective evidence has suggested that a decline in the lymphocyte count may be predictive of mortality irrespective of the individual's age or type of disease.

Specific studies, particularly those focused on samples under stress, find that interventions that include some form of social support have enhancing effects on natural killer (NK) cell cytotoxicity, lymphocyte proliferation, and cell-mediated immunity, as indexed by diminished herpes simplex virus (HSV) antibody titers. These effects would be expected if social support could enhance immune function at least for individuals under stress. It is also noteworthy that massage, which requires physical contact between at least two individuals, is also effective in producing immune-enhancing effects such as increased NK cell activity.

IV. EXPRESSION OF EMOTION

The idea that emotions, particularly negative emotions, influence one's health has a long history. It has been thought that the tendency to experience negative emotions and the failure to recognize or communicate such emotions may be particularly detrimental.

Research conducted over the last 15 years has documented that disclosure of traumatic events, a source of negative emotion, appears to be associated with better health and alters autonomic activity and immune function. Pennebaker has been a lead investigator in this area. His work and that of others have shown that when individuals communicate events that produce negative emotional states, they exhibit elevations in NK cell activity and lymphocyte proliferation. Lower HSV titers and stronger responses to immunization are found. Changes in the leukocyte count have also been observed. Overall, lymphocyte numbers may increase, but various lymphocyte subtypes tend to exhibit different responses. For instance, helper T lymphocytes appear to decrease. However, because studies in this area are limited, caution is in order regarding the robustness of these observations.

It is noteworthy that communication of traumatic events requires verbal processing, and for most individuals this means left hemisphere brain activity. Such activity could serve to counteract right hemisphere activity and the autonomic outflow capable of downregulating immune responsiveness. Also interesting is that social engagement makes it more likely that individuals will communicate verbally and possibly even disclose troubling events or circumstances. Consequently, social engagement may be capable of sustaining health by promoting verbal communication and the disclosure of negative experiences in particular.

V. SLEEP AND RELAXATION

Sleep is most fundamental to the recuperative processes of living organisms. Humans spend nearly one-third of their life sleeping. Metabolic activity associated with energy usage tends to promote sleep. For instance, adenosine accumulation due to energy expenditure promotes sleep. Blockade of adenosine receptors by substances such as caffeine sustains wakefulness.

Acute infection induces slow-wave sleep as part of the mobilization of resources aimed at fighting infection. If slow-wave sleep does not develop in response to infection, the severity of the clinical disease appears to be augmented. Moreover, when pathogens persist within the organism, sleep becomes disrupted. Such disruption is also a characteristic of chronic inflammatory disorders such as rheumatoid arthritis (RA), fibromyalgia, and chronic fatigue syndrome. In the case of RA, pain appears to be aggravated by the sleep disturbance, contrary to what is sometimes assumed to occur—namely, that pain is the cause of sleep disturbance. Inflammatory conditions are actually associated with fragmented sleep, with frequent arousal to the point of near wakefulness over the course of the night. Sleep disturbance is also evident in major depression, which has been found to be associated with proinflammatory activity.

In essence, it appears as if regular sound sleep is essential to the functioning of the immune system. Sleep is induced during the acute phase of an infectious disorder, presumably in an effort to enhance the defensive response of the host. However, under conditions of persistent infection or chronic inflammation, sleep seems to be disturbed, possibly in an attempt to turn off immune activity, but which, in turn, impairs the individual's ability to mount effective immune responses to novel challenges.

Wakeful relaxation has been thought to have health benefits because it is assumed to counteract the effects of stress. Studies of relaxation effects on immune function have examined various techniques of induction, including those relying on mental exercises and biofeedback. The induction of relaxation, by any means, is associated with decreases in negative affect and sympathetic nervous system activity and produces alterations of hormone levels in the circulation. Consequently, effects on immune activity would be expected. However, the available evidence provides few consistent effects on measures such as leukocyte counts, lymphocyte proliferation, or NK cell activity. Nonetheless, individual studies have shown that relaxation is associated with decreased leukocyte counts. In older subjects, relaxation training serves to enhance NK cell activity, and in the case of medical students who regularly practice relaxation, decreases in immune function during periods of examination are attenuated. The most consistent finding associated with relaxation is an increase in total secretary IgA, noteworthy, because hypnosis, which is thought to induce relaxation, also produces reliable increases in secretory IgA.

VI. EXERCISE AND PHYSICAL ACTIVITY

Exercise and physical activity result in alterations in bodily systems including the immune system. The exertion associated with strenuous physical activity produces cardiovascular, respiratory, endocrine, and metabolic changes, all of which can contribute to the observed alterations in immune function. Furthermore, exercise can cause physical trauma and induce inflammation.

The immune changes produced by exercise are similar to the changes caused by psychosocial stress. However, in contrast to the adverse effects of some forms of psychosocial stress, repetitive physical exercise is associated overall with better health and longevity. A notable exception is the frequently observed susceptibility to upper respiratory tract viral infection after bouts of intense exercise, particularly in the context of competition.

A. LEUKOCYTES

For more than 100 years, it has been known that acute exercise causes a transient increase in the white blood cell count (leukocytosis). Hemodynamic changes (i.e., increased heart rate and stroke volume) may in part cause the leukocytosis. Hormonal changes associated with strenuous exercise may also contribute to the increased leukocyte count, particularly the exercise-induced increases in the levels of epinephrine, norepinephrine, gonadal steroids, cortico-steroids, β-endorphin, and growth hormone.

As noted above, the best-documented effect of exercise is leukocytosis, with neutrophils being largely responsible for the effect. The effect is transient, with levels returning to preexercise baseline within 24 hours, even if exercise is maintained. Despite increased numbers in the circulation, the adherence capacity of neutrophils and their bactericidal capability are diminished after exercise. Bactericidal activity may be transiently increased but then gives way to a more lasting decrease. In contrast, neutrophil chemotaxis and phagocytosis remain unaltered.

Mononuclear phagocytes (monocytes) are also increased in the circulation after exercise, but the effect is transient, lasting no more than 2 hours. Enhanced phagocytic activity has been observed, but this may be tissue specific in the sense that mononuclear cells from different tissues show differential activity. Specifically, cells derived from the airway appear more active than those from the peritoneal cavity.

Lymphocytes increase rapidly in the circulation after exercise begins. All lymphocyte subtypes are increased, though to differing degrees. Lymphocyte numbers return to baseline quickly after the cessation of exercise. Hormonal influences appear prominent in that epinephrine enhances lymphocytosis, whereas cortisol counteracts and can even block the effect of exercise on the lymphocyte count. The largest increase is observed in the NK cells, followed by $CD8^+$, $CD4^+$, and B cells. NK cells have the highest number of β-adrenergic receptors, followed by $CD8^+$ and then by $CD4^+$ cells, an observation that further supports the role of adrenergic stimulation in the production of lymphocytosis.

NK cell cytotoxicity is increased shortly after exercise possibly because of increased levels of β-endorphin but gradually declines even if exercise persists. The $CD8^+$ increase seems mostly mediated by "memory" T cells. T-cell proliferation in response to mitogens is decreased during and for sometime after the cessation of exercise. Delayed-type hypersensitivity is also decreased after exercise. In contrast, B-cell proliferation appears enhanced after exercise.

B. Acute-Phase Proteins, Antibodies, and Cytokines

The evidence further documents that exercise induces the production of acute-phase proteins such as complement components. There are also changes in antibody and cytokine levels. Intense exercise is associated with an increase in some complement components (e.g., C3 and C4). There does not appear to be any effect on immunoglobulin levels in the circulation. However, secretory sIgA tends to be decreased after intense exercise but returns to normal rapidly. The effect on sIgA may be secondary to diminished secretions from mucosal tissues.

Cytokines, especially those involved in promoting inflammation, are increased in response to exercise. Early studies demonstrated that interferon is increased by exercise. Interleukin-1 (IL-1) was also found to be increased by exercise, but other work has shown increased IL-1RA as well, which would be expected to neutralize the effect of IL-1. Increased IL-6 has been consistently observed by other studies, along with small increases in IL-1β and tumor necrosis factor-α (TNF-α). There appears to be an increase in both proinflammatory cytokines and anti-inflammatory cytokines such as IL-10. Other forms of physical activity such as the practice of yoga appear to have anti-inflammatory endocrine effects. For instance, routine practice of yoga has been associated with high cortisol levels but low catecholamine levels in the circulation.

Thus, in many ways the response to exercise is similar to that seen after trauma and infection and, as previously noted, is similar to that due to psychosocial stress. In effect, exercise may be seen as a form of self-imposed stress. It leads to transient changes in immune function, such as those that result from stress, but exercise is not associated with disease and in fact it tends to be associated with less morbidity and enhanced longevity.

VII. NUTRITION

The immune system, as is true for all other bodily systems, is influenced by the nutritional state of the individual. In this regard, it is important to keep in mind that the nutritional abundance of modern affluent societies is not the standard for determining what is adequate for immune function. The immune system evolved and established its role in the protection of the organism when sources of nutrition were not as readily available.

Nutritional research, most based on animal studies but also conducted in parts of the world where human undernourishment is widespread, has amply documented the role of caloric intake and specific nutrients in immune competence. Overall, that evidence shows that most nutritional deficits are associated with suppressed immune responsiveness. However, some restriction of caloric intake has been found to make organisms more resistant to disease. Moreover, anorexia is part of the sickness response aimed at combating infection. Especially noteworthy is the observation that excessive consumption of food tends to be associated with immune suppression and is known to increase the risk of cancer.

The impact of nutrition begins *in utero* in ways that bias how the organism functions throughout the life span. The role of early nutritional excesses is not well understood. However, obesity is a risk factor for various immunological disorders.

Evidently, "more" is not better with respect to immune function, and possibly even "average" in an environment of abundance may hamper immune function.

A. MALNUTRITION

Malnutrition from low-protein, low-calorie diets causes a state of immuno-deficiency. There is atrophy of the lymphoid tissues, lymphocyte counts are low, complement components are diminished, and both cellular and humoral immune responses are reduced. In contrast, leukocyte counts and immunoglobulin levels remain unchanged or increase. Despite low lymphocyte numbers, the leukocyte count increase is largely due to an increase in the number of neutrophils. This is a familiar pattern seen in response to stress, and obviously starvation classifies as stress. However, as noted earlier in this chapter, moderate lowering of caloric intake tends to be associated with increased T-cell responses. In fact, leanness seems to protect against infection and cancer. Nonetheless, the increased immune re-sponsiveness noted during mild anorexia cannot be sustained if body weight falls below 60% of the ideal for the individual. Furthermore, the benefits of caloric restriction also depend on aspects of the diet such as the source of calories (e.g., lipids vs protein) and the presence of other essential nutrients (i.e., vitamins and minerals). The evidence about diet and immune function is clear that malnutrition is detrimental. Overnutrition seems to have adverse effects as well. Moderate nutrition, if it is balanced, appears to be best. However, the questions of what is optimal and whether supplementation with specific nutrients would further en-hance immune function remain without clear-cut answers. This is an area in which characteristics of the individual, such as age and health status, must be taken into consideration.

B. INDIVIDUAL FACTORS AND NONLINEAR EFFECTS

The need to consider individual factors is immediately evident when one examines the findings on the effect of specific nutrients on immune function. *Nucleic acids* are necessary in the diet to maintain cell-mediated immunity, but supplementation seems beneficial only in infants or in individuals recovering from sepsis. *Amino acids* are essential for the overall function of the immune system. Arginine and glutamine appear to have some value as supplements, even over and above the effect of a balanced diet. *Minerals* are also crucial to the function of the immune system, with deficiencies being detrimental across the board. However, supplementation is not necessarily beneficial and may even be harmful. Copper interferes with phagocyte function at low and high concentrations. Iron and zinc, though essential to immune function, decrease in the blood as part of the acute response to infection. In effect, they are made less available to pathogens, so supplementation could be detrimental by benefiting pathogens. Moreover, zinc supplementation can interfere with the absorption of copper and thus create a deficit. Magnesium has a dampening effect on inflammation and so low levels may promote inflammatory processes to a degree that could be detrimental. By the same token, high levels could interfere with the protective aspects of adequate inflammation. Selenium is the only mineral that seems to have some immune-enhancing effects at higher doses, an effect that may be attributable to its antioxidant properties.

C. ANTIOXIDANTS: VITAMINS AND MINERALS

Antioxidants are compounds capable of neutralizing highly reactive oxygen-free radical molecules, also known as "reactive oxygen species," normally produced during cellular metabolism or by the immune system in the process of destroying microorganisms. Such oxygen-free radicals are capable of damaging cell membranes and the cell's DNA. The body has a limited ability to neutralize such molecules. Thus, a major source of antioxidants comes from the diet. As noted earlier in this chapter, selenium is an antioxidant, as are some vitamins and other constituents of the diet, all of which play a role in preventing tissue damage by neutralizing oxygen-free radicals.

Vitamin deficiency impairs immune function, but again, as in the case of minerals, supplementation may have negative effects. Vitamin A has been called the "anti-infective vitamin." It is necessary to sustain the functional integrity of innate defenses, which serve to block pathogen penetration. Vitamin A is also crucial to the mounting of specific immune responses. However, supplementation with vitamin A has been shown to benefit only undernourished populations. B-complex vitamins are only detrimental when deficient. However, vitamin D can downregulate cell-mediated responses when present in excess. Vitamins C and E are antioxidants and can enhance immune responses, particularly in elderly populations. They both can have adverse effects in excess. For instance, vitamin E at high doses (more than 1600 mg per day) can hinder neutrophil phagocytosis.

The immune-enhancing effects of antioxidants such as selenium and vitamins C and E may be related to the blocking action of oxygen-free radicals on the function of Th1 lymphocytes. The latter are suppressed by oxygen-free radicals because they promote the production of such oxygen reactive species with the potential to cause tissue damage. Thus, antioxidants may promote cell-mediated immunity by diminishing negative feedback on Th1 lymphocytes.

Beta-carotene, a constituent of plant foods, also functions as an antioxidant and has been found to protect against immunosuppression induced by reactive oxygen species. Beta-carotene consumption is associated with a lower incidence of cancer, perhaps because of the strong possibility that oxygen-free radicals, either endogenously generated or from environmental sources (e.g., ozone, tobacco smoke, and ultraviolet radiation), are carcinogenic. However, the situation is complicated by the possibility that antioxidants may protect cancer cells once they are produced. In other words, antioxidants may be beneficial only in a preventive mode but not as a component of treatment.

D. Lipids

Lipids in the diet are necessary, but when consumed in excess (more than 40% of total energy intake) they tend to be associated with immune suppression. Essential fatty acids ($\omega 3$ and $\omega 6$) are derived from the diet and are the building blocks for the eicosanoids such as the prostaglandins (e.g., prostaglandin $E_2[PGE_2]$) and the leukotrienes (e.g., leukotriene $B_4[LTB_4]$), molecules that promote inflammation when released from mast cells. Eicosanoids produced from $\omega 3$-fatty acids, obtained predominantly from fish oils, are less proinflammatory than those derived from $\omega 6$-fatty acids. The latter may promote inflammation to a degree that could be problematic. Thus, the amount and profile of lipid ingestion could be a factor in some autoimmune disorders and cardiovascular disease, in which inflammation is seen as one aspect of the underlying pathophysiology. Similarly, malignancy arising from the carcinogenic potential of chronic inflammation could be traced to lipid ingestion at least in some cases.

E. Modulation by Hormones

Finally, the availability of nutrients to various tissues is subject to modulation by hormones. For instance, insulin is released in response to food ingestion and promotes the storage of nutrients. Thyroid hormones, in contrast, attenuate glucose storage, decrease total body lipids, and diminish protein synthesis. The classic stress hormones (i.e., glucocorticoids and catecholamines) similarly promote glucose release from storage and lipid breakdown and inhibit protein synthesis in muscle but not in the liver. Growth hormone has biphasic effects on glucose and lipids, initially decreasing serum levels and then increasing them while promoting protein synthesis. Gonadal steroids increase blood glucose, decrease fat deposits, and enhance protein synthesis in muscle tissue. Cytokines (e.g., IL-1β and TNF-α) and eicosanoids (e.g., PGE_2 and LTB_4) have effects similar to those of stress hormones. Essentially, they produce a redistribution of nutrients, moving them from storage sites to the tissues most directly involved in the response to specific threats or challenges.

Enhancement of immune function via nutrition appears to be best accomplished by a balanced diet that does not rely heavily on fat for energy (i.e., no more than 35% of total energy), and that is not deficient of ω3-fatty acids. Supplementation of a balanced diet does not appear beneficial except in older individuals or during specific phases of immune activity or prolonged periods of immune dysregulation.

VIII. CONCLUDING COMMENTS

This chapter has focused on activities that have been consistently found to be associated with health maintenance. These include positive expectations, appropriate expression of negative emotion, and social engagement in conjunction with balanced nutrition, physical activity, and rest. It has been assumed that benefits may at least in part derive from changes that counteract or buffer the impact of stress on the immune system.

Evidence from studies of the placebo effect, hypnotic suggestion, and classical conditioning all point to the influence of expectation, be it conscious and unconscious, over processes that can promote healing and improve well-being. This evidence serves to document that patterns of brain activity, at least for some individuals, can facilitate changes in peripheral tissues that initiate healing or inhibit tissue-damaging activity.

The evidence further suggests that social engagement may be beneficial because it creates the opportunity to verbally communicate feelings and experiences that are troubling the individual. For most individuals, such verbal expression would activate the language-processing networks in the left cerebral hemisphere, a shift that could in turn modulate right cerebral hemisphere activity in ways that modify autonomic outflow and thereby enhance immune responsiveness.

Recuperation in the form of sleep and relaxation naturally would be expected to be beneficial. It is of interest that acute infection promotes sleep in conjunction with immune activation. In turn, immune overactivation in the form of chronic inflammation appears to disrupt sleep, conceivably in an effort to shut off overactivation. At least in some individuals under stressful situations, wakeful relaxation appears to facilitate immune responsiveness.

A routine of physical exercise is associated overall with better health and longevity. Regular exercise is best viewed as self-regulated chronic stress. The immune changes associated with exercise are similar to those seen in response to acute psychosocial stress. In many respects the protective influence of exercise is reminiscent of the ways in which chronic psychosocial stress may toughen some individuals against adversity. It also brings to mind that training the body to deal with challenges makes the individual better prepared to cope with unexpected challenges.

Finally, the role of nutrition is clearly not a simple matter of abundance. Severe nutritional deficits produce immune deficiencies similar to the effects seen in response to other forms of stress. Overconsumption also tends to be associated with decreased immune responsiveness and is known to increase the risk of disease. Overconsumption may constitute a form of stress. The evidence indicates that caloric restriction accompanied by body leanness enhances resistance to disease and prolongs life. Indeed, anorexia is a component of the acute-phase response to infection. Moreover, the composition of the diet is a factor that contributes to the neutralization of products of oxidative metabolism, provides the building blocks for making and regulating key enzymes, and dampens inflammatory responses. In addition, modulation of nutrient availability via adjustments in ingestive behavior and metabolic activity can inhibit the growth of pathogens.

Overall, it seems fair to conclude that although the evidence is limited and not entirely consistent, important insights can be gleaned nonetheless even if strong conclusions are not warranted regarding immune enhancement. Positive expectations appear to be important, but they may need to be set in motion in the context of a credible healing relationship and with the assistance of a remedy or healing ritual. Being with others and using the opportunity to communicate what troubles the individual may counteract the negative effects of sustained distress on the body. Adequate rest, diet, and exercise are probably important as background factors that allow the individuals to maintain a readiness to respond effectively to challenges and, thus, play a preventive role. These are not new insights; they are clearly put forth in ancient writings concerned with longevity. What has dramatically changed is our understanding of the likely mechanisms underlying the beneficial effects of the factors discussed in this chapter.

IX. SOURCES

Berkman, L.F., Glass, T., Brissette, I., & Seeman, T.E. (2000), "From Social Integration to Health: Durkheim in the New Millennium," *Social Science & Medicine*, 51, pp. 843–857.

Booth, R.J., Petrie, K.J., & Pennebacker, J.W. (1996), "Changes in Circulating Lymphocyte Numbers following Emotional Disclosure: Evidence of Buffering?" *Stress Medicine*, 43, pp. 293–306.

Brady, H. & Brady, D. (2000), *The Placebo Response*, New York, HarperCollins.

Christensen, A.J., Edwards, D.L., Wiebe, J.S., et al. (1996), "Effect of Verbal Self-Disclosure on Natural Killer Cell Activity: Moderating Influence of Cynical Hostility," *Psychosomatic Medicine*, 58, pp. 150–155.

Fawzy, F.I., Kemeny, M.E., Fawzy, N.W., et al. (1990), "A Structured Psychiatric Intervention for Cancer Patients: Changes Overtime in Immunological Measures," *Archives of General Psychiatry*, 47, pp. 729–735.

Green, M.L., Green, R.G., & Santoro, W. (1998), "Daily Relaxation Modifies Serum and Salivary Immunoglobulin and Psychophysiologic Symptoms Severity," *Biofeedback and Self-Regulation*, 13, pp. 187–199.

Hall, N.R.S. & O'Grady, M. (1991), "Psychosocial interventions and immune function," in Ader, R., Felten, D.L., & Cohen, N. (eds) *Psychoneuroimmunology*, 2nd ed, San Diego, Academic Press.

Halley, F.M. (1991), "Self-Regulation of the Immune System through Biobehavioral Strategies," *Biofeedback and Self-Regulation*, 16, pp. 55–74.

Harrington, A. (1997), *The Placebo Effect: An Interdisciplinary Exploration*, Cambridge, Mass, Harvard University Press.

Hewson-Bower, B. & Drummond, P.D. (1996), "Secretary Immunoglobulin A Increases during Relaxation in Children with and without Recurrent Respiratory Tract Infections," *Journal of Developmental and Behavioral Pediatrics*, 17, pp. 311–316.

Hoffman-Goetz, L. & Pedersen, B.K. (2001), "Immune responses to acute exercise: Hemodynamic, hormonal, and cytokine influences," in Ader, R., Felten, D.L., & Cohen, N. (eds) *Psychoneuroimmunology*, 3rd ed, volume 2, San Diego, Academic Press.

Hughes, D. (2002), "Nutritional effects on human immune function," in Brostoff, J. & Challacombe, S.J. (eds) *Food Allergy and Intolerance*, 2nd ed, London, WB Saunders.

Ironson, G., Field, T., Scafidi, F., et al. (1996), "Massage Therapy Is Associated with Enhancement of the Immune System's Cytotoxic Capacity," *International Journal of Neuroscience*, 84, pp. 205–217.

Johnson, V.C., Walker, L.G., Heys, S.D., et al. (1996), "Can Relaxation Training and Hypnotherapy Modify the Immune Response to Stress and Is Hypnotizability Relevant?" *Contemporary Hypnosis*, 13, pp. 100–108.

Kiecolt-Glaser, J.K., Glaser, R., Strain, E.C., et al. (1986), "Modulation of Cellular Immunity in Medical Students," *Journal of Behavioral Medicine*, 9, pp. 5–21.

Kiecolt-Glaser, J.K., Glaser, R., Williger, D., et al. (1985), "Psychosocial Enhancement of Immuno-Competence in a Geriatric Population," *Health Psychology*, 4, pp. 25–41.

Klopfer, B. (1957), "Psychological Variables in Human Cancer," *Journal of Projective Techniques*, 21, pp. 331–340.

Krueger, J.M., Fang, J., & Majde, J.A. (2001), "Sleep in health and disease," in Ader, R., Felten, D.L., & Cohen, N. (eds) *Psychoneuroimmunology*, 3rd ed, vol 1, San Diego, Academic Press.

McGrady, A., Conran, P., Dickey, D., et al. (1992), "The Effects of Biofeedback-Assisted Relaxation on Cell-Mediated Immunity Cortisol, and White Blood Cell Count in Healthy Adult Subjects," *Journal of Behavioral Medicine*, 15, pp. 343–354.

Melmed, R.N. (2001), *Mind, Body, and Medicine: An Integrative Text*, New York, Oxford Press.

Miller, G.E. & Cohen, S. (2001), "Psychosocial Interventions and the Immune System: A Meta-analytic Review and Critique," *Health Psychology*, 20, pp. 47–63.

Peavey, B.S., Lawlis, G.F., & Goven, A. (1985), "Biofeedback Assisted Relaxation: Effects on Phagocytic Capacity," *Biofeedback and Self-Regulation*, 10, pp. 33–47.

Pennebacker, J.W. (1995). *Emotion, Disclosure, and Health*, Washington, DC, American Psychological Association Press.

Pennebacker, J.W., Kiecolt-Glaser, J.K., & Glaser, R. (1988), "Disclosures of Traumas and Immune Function: Health Implications for Psychotherapy," *Journal of Consulting and Clinical Psychology*, 56, pp. 239–245.

Simon, H.B. (1991), "Exercise and human immune function," in Ader, R., Felten, D.L., & Cohen, N. (eds) *Psychoneuroimmunology*, 2nd ed, San Diego, Academic Press.

Sloan, R.P., Bagiella, E., & Powell, T. (1999), "Religion, Spirituality and Medicine," *Lancet*, 353, pp. 664–667.

Udupa, K.N. & Singh, R.H. (1979), "Yoga in relation to the brain pituitary adrenocortical axis," in Jones, M.T., Gillham, B., Dallman, M.F., & Chattopadhyay, S. (eds) *Interaction within the Brain–Pituitary Adrenocortical System*, London, Academic Press, pp. 273–278.

Wegner, D.M. (2002), *The Illusion of Conscious Will*, Cambridge, Mass, MIT Press.

Yoshida, S.H., Keen, C.L., Ansari, A.A., & Gershwin, M.E. (1999), "Nutrition and the immune system," in Shils, M.E., Olson, J.A., Shike, M., & Ross, A.C. (eds) *Modern Nutrition in Health and Disease*, 9th ed, Baltimore, Williams & Wilkins, pp. 725–750.

Zachariae, R. (2001), "Hypnosis and immunity," in Ader, R., Felten, D.L., & Cohen, R. (eds) *Psychoneuroimmunology*, 3rd ed, vol 2, San Diego, Academic Press.

Zeitlin, D., Keller, S.E., Shiflett, S.C., et al. (2000), "Immunological Effects of Massage Therapy during Academic Stress," *Psychosomatic Medicine*, 62, pp. 83–84.

CHAPTER 12

Integration and Implications

I. Introduction   225
II. Synopsis   225
III. Microenvironments and Complexity   228
IV. Unnecessary and Insufficient   228
V. Implications for Research   229
    A. Limits of Reductionism   229
    B. Major Dimensions of Health
       State Space   229
VI. Implications for Health Care   232
    A. Holistic-expanded Perspective   232
    B. Individualized Health Education   232
VII. Economic Considerations   233
VIII. Concluding Comments   233
IX. Sources   234

I. INTRODUCTION

This book has delved into basic details of how the immune system operates and how it is modulated by neuroendocrine signals partially in response to psychosocial stress. The role of psychosocial stress in major forms of disease and the impact of immune activity on psychosocial function have been reviewed. Consideration has also been given to practices that may buffer the impact of psychosocial stress on health. This chapter highlights the key insights derived from psychoneuroimmunology and draws implications for research and health care.

II. SYNOPSIS

Psychoneuroimmunology is a higher order scientific discipline that has emerged from reductionistic approaches that isolate bodily systems or their components to characterize aspects of their operation but that consequently fall short of elucidating how the systems operate in their natural context. Psychoneuroimmunology begins to address the needed integration and grapple with the level of biological complexity necessary to understand health. In a sense, psychoneuroimmunology reasserts early ideas about the fundamental unity of the organism and the notion that health rests on proper balance, but with a modern understanding of the systems that must be in balance.

The immune system operates at the boundary between the organism and other life forms. It is composed of elements that protect the integrity of the organism by blocking, neutralizing, or destroying microscopic predators or aberrant cells originating within the organism. Clearly, a balance must be struck between the need to defend against pathogens and the need to safeguard healthy tissue. The unpredictability of pathogen characteristics over time requires that the immune system remain poised to deal with novelty while retaining proven responses to handle predictable or more stable aspects of the microbial world. Thus, the immune system embodies components that innately respond to pathogens and those that are capable of detecting pathogens with novel characteristics. In addition, not all individuals are equally prepared to deal with a given pathogen at a particular time, because at any given moment the specificity of antibodies and T-cell receptors available to a given individual can only be a subset of what is possible and may not be optimal to neutralize a given pathogen. However, across a human population, there would likely be sufficient readiness to withstand novel challenges by pathogens. A related phenomenon is that there are changes with age in such readiness, being lowest during early life and again later in life. The balancing act between self and nonself and the overall readiness of a person's immune defenses are further subject to influences mediated by the endocrine system and the peripheral nervous system (PNS) in response to other changes in the individual's environment at large.

The endocrine and immune systems intercommunicate to achieve balanced activity compatible with life given specific contexts. Growth-oriented hormones that modulate the storage and use of nutrients have effects on immune function by how available resources are allocated. In particular, the pituitary–adrenal axis responds to changes in context that mobilize metabolic resources to confront a perceived threat or challenge. This can result in temporary suppression or potentiation of immune function. Procreation, the most life-sustaining activity, adds another layer of regulation to the immune system, because successful pregnancy requires tolerance of foreign tissue and hinges on timely suppression of immune attack. In essence, the endocrine system can be viewed as a source of signals that shift the balance of immune activity in ways that sustain individuals and enhance the chances that they will produce offspring.

Endocrine modulation of immune activity tends to be global in that hormones circulate throughout the body. Some degree of specificity can be attained by the expression of hormone receptors in specific cell types or tissue compartments. However, the localization of influence is greatly enhanced via the PNS. The PNS provides the structural basis for sensing and modifying activity within organismic microenvironments, roughly coinciding with the territory innervated by specific PNS terminals. The activity of PNS terminals within specific microenvironments is subject to various influences including those originating within higher order central nervous system (CNS) structures. The latter is simply a consequence of the interconnectivity that is abundantly evident. In essence, immune cells operate under the influence of microenvironmental conditions that partially reflect hormonal secretions and local neural activity.

The notion of stress captures the fact that contextual change is a major force that creates disequilibrium within the bodily systems responsible for sustaining life. It is during such moments of transition that both vulnerability to disease and the opportunity for robustness enter the picture and ultimately become manifest by how the organism is able to respond to the challenge. Mounting an effective response is a form of discovery; it is not predictable, but its occurrence renders the individual better prepared to handle similar forms of future adversity.

Psychosocial stress has neuroendocrine and immune effects in humans. These effects must begin within the CNS, and then they are transmitted to the endocrine system and the PNS. Studies have typically shown a biphasic response. Initially hormone levels in the blood are increased and then return to baseline or become diminished. A similar dynamic is evident in autonomic nervous system activity. This is expected in systems with negative feedback. Peripheral neuroendocrine consequences of psychosocially induced CNS activity feed back to modulate CNS activity and thereby influence the impact of future psychosocial inputs. Brain structures such as the hippocampus appear as key sites wherein various influences coverage and alter neurons with effects on mental and neuroendocrine responses. These transactions, in turn, have an impact on immune activity. There is an initial mobilization of leukocytes, often accompanied by a dampening of their activity. Antibody titers tend to be increased in the circulation, but this appears to result from decreased secretion onto mucosal surfaces or occurs secondary to diminished cell-mediated immune activity. These findings suggest that at least temporarily the immune system is less able to deal with new challenges. Nonetheless, such effects are detected on average with the norm across individuals being a marked variability of immune reactivity in response to stress.

Disease is a form of stress and begins within microenvironments. This is particularly true about cancer but is also consistent with the initially local actions of infectious agents, allergens, and autoimmune responses. What occurs at the level of microenvironments is a complex function of external irritants and the pattern of gene expression in the cells populating the region. What transpires is further influenced by the nutritional state of the organism and signals arising from the endocrine glands and the local innervation by the PNS. Psychosocial factors exert some of their influence by modulating the neural and endocrine signals at the microenvironmental level. The specific configuration of such variables sets the stage for the likelihood that a localized pathological change will occur and expand sufficiently to cause clinical symptoms of disease.

Diseases appear most compatible with complex models of causality. This is equally the case with respect to psychiatric disorders. Such disorders cannot be simply attributed to social experience interacting with genetic susceptibility. There is sufficient evidence to suggest that infectious agents and toxic substances can play a role in the genesis of psychiatric disorders. With respect to the latter, even some of the medicines used to control psychiatric symptoms may create changes that further aggravate the disorder. This is similar to what has been suggested in the case of medicines used to treat allergic symptoms.

Finally, turning to what seems to help individuals maintain health, one encounters a familiar array of practices, most of which have been recognized since antiquity. They include mental attitude in the form of positive beliefs and expectations, as well as social engagement, which allows for physical contact and creates the conditions for channeling emotional reactions into verbal communications. Also evident is that appropriate relaxation, physical activity, and adequate nutrition all place the individual in an advantageous position with respect to health maintenance. In other words, over a lifetime individuals who are able to adhere to these practices spend much less time in disease states.

III. MICROENVIRONMENTS AND COMPLEXITY

Psychoneuroimmunology underscores the notion that disease begins within microenvironments. A microenvironment can be defined at different levels, such as that immediately adjacent to a segment of DNA containing a single gene or that surrounding a single cell or a small volume of tissue composed of a few interacting cells. In fact, one could go even below the single-gene level or certainly define a small volume of tissue to include more than a few interacting cells. Essentially, the number of possible definitions is quite large. However, by following the flow of activity into progressively more circumscribed microenvironments, one gains the important realization that there is no corresponding decrease in the complexity of regulatory influences. In effect, as one moves across levels or progressively restricts the focus of interest, one discovers a conservation of complexity.

IV. UNNECESSARY AND INSUFFICIENT

Microenvironments are the sites wherein aberrant activity first appears, but for disease to occur such activity must be sustained and expand to encompass a sufficiently large territory so that organ or system function is compromised. Any and all the influences capable of blocking the stabilization and expansion of such activity became crucial to sustaining health.

Many influences or variables have been implicated in health maintenance. The evidence also raises the possibility that a variable capable of having an impact on health is probably best viewed in a given case as neither necessary nor sufficient to bring on disease. Generally speaking, influential variables come in two varieties: those that increase the probability of disease and those that decrease the probability. However, this is not entirely an intrinsic property of the variables; it depends on which other variables are in the mix. What appears likely is that if there is an imbalance of sufficient magnitude so that the facilitating influences predominate, the disease process will be set in motion.

V. IMPLICATIONS FOR RESEARCH

A. LIMITS OF REDUCTIONISM

Recognition that complexity cannot be avoided by progressively narrowing the focus, a notion that has been dubbed "self-similarity" by chaos theorists, along with the observation that a specific influence can act to inhibit or excite a given process depending on the mix of other concurrent influences, has profound implications for research.

Reductionistic approaches must be viewed as necessary first steps but ultimately inadequate to understand how a complex system at any level operates in its natural context. This is because the level of control necessary for a good experiment creates an artificial situation, giving rise to outcomes that are highly dependent on the conditions set by the experiment. It is not possible to be sure about what would be observed if the experiment were slightly altered. In fact, even very careful attempts at replicating an experiment can yield discrepant results. Scientific methodology is not without limitations and it is worth remembering that the most precise sciences are those that deal with inanimate objects. Once an entity acquires the characteristics that define life, its ability to adapt and evolve makes it difficult to predict.

Efforts to understand disease by reductionistic approaches, which seek to demonstrate the controlling variables for all time, will continually fall short. This is because organisms are complex systems, they continually alter their own context, and thus, they modulate the influence of other variables. As a result, even the best constructed experiments will fail to converge often enough to create uncertainty. Of course, this is not an argument for giving up on such research; rather it is a call to acknowledge the limitations of efforts to established fundamental causes of disease.

B. MAJOR DIMENSIONS OF HEALTH STATE SPACE

The type of research necessary to do justice to the complex nature of disease will need to track individuals over time within a psychoneuroimmunological framework defined by variables that quantify the dimensions that influence health.

Disease states appear as regions within a multidimensional health space. Within such a framework, *prevention* refers to actions that steer individuals away from regions that define disease states, and *treatment* corresponds to interventions that lead individuals out of disease state regions.

Age and gender are linked to disease probabilities and thus quantify important dimensions of health state space. Many disorders exhibit gender biases or become increasingly probable as one ages. Moreover, now that the human genome has been systematically catalogued, it will become increasingly possible to detect specific genes or genetic profiles that increase the probability of some disorders. However, two issues must be kept in mind. The first is that even if one takes into consideration the estimated 30,000 genes in the human genome, it would still be impossible to approach certainty with respect to any given disease occurring at a particular time. Essentially, all the relevant information is not present in the genome and key determinants may not even materialize until the moment of disease onset, much like the proverbial straw that breaks the camel's back. The second concerns the possibility that a genetic profile that increases risk for a particular disorder may simultaneously decrease risk for a variety of other disorders that tend to appear within the same age-group. Obviously, genetic profiles that limit longevity cannot be regarded as protective of individuals with respect to diseases of old age. Similarly, genetic profiles that are protective against infectious diseases and early forms of malignancy may be associated with later onset chronic diseases, simply because they promote longevity, but should not be regarded as specific risk markers.

A major group of variables may be subsumed under the rubric of exposure. The possibilities in this regard are vast and the effects variable depending on dosage, duration of exposure, and individual characteristics. Human-made chemicals, including pharmaceuticals, are part of this influence, which also includes climate, various forms of radiation, contaminants in air and water, and diet and further encompasses recreational or work-related exposures. Essentially, one lives in a minispace with a particular set of physicochemical properties, which change overtime by how humans expand the molecular composition of nature via synthetic chemistry and technology. Understanding the dynamics of disease requires a characterization of exposure.

Already, one should have concerns about how one could track so many variables, even though not all the relevant ones have been cited yet. This is in fact the major hurdle to fundamentally understanding disease. It is clear that both individual characteristics as revealed by psychoneuroimmunology and exposure must be incorporated into any model of health. However, individual characteristics extend well beyond the realms of age, gender, and genetics. It is also necessary to contend with the function of organs such as the liver or the kidneys, because as is true for the brain or the immune system, not all of what transpires at the organ level is fully specified by the relevant genes. However, given the focus of this book the relevance of the brain and the immune system is emphasized. With respect to the immune system, individuals need to be represented in terms of Human Leukocyte Antigen (HLA) types. Functional indices of immune responsiveness such as the reactivity of leukocyte types to challenges with foreign antigen or other activators must be quantified. In the case of the brain, it is important to obtain direct measures of brain activity, assess behavior, and mental phenomena. At this juncture, it may suffice to represent individuals in terms of psychological characteristics and typical patterns of behavior. Finally, the extent to which individuals are socially integrated adds yet another dimension to the framework for assessing risk of diseases or determining whether they are on a trajectory leading to disease regions of health state space.

Finally, as it seeks its own survival, the microbial world becomes a major force moving individuals into disease states. Thus, to have a precise estimate of the probability of disease, one must gauge exposure to microbes. It further becomes necessary to anticipate that the microbial world is not fixed but exhibits the ability to evolve rapidly. As is now well established, human attempts to eliminate pathogens have become a force in their evolution, and consequently unanticipated assaults on the health of humans have occurred. Unless this issue is seriously addressed, the search for antimicrobial agents will simply perpetuate their tendency to evolve in ways that evade our attempts to poison them.

The preceding considerations lead to the conclusion that an individual's state of health at any time is a complex function of age, genetic characteristics, and specific organ or system dynamics (including those that give rise to mental life, behavior, and social interaction) in response to challenges posed by exposure to the physicochemical and microbial aspects of the individual's immediate context. Adequate characterization of all these factors greatly exceeds the current methodology, even in the most sophisticated research.

In order to capture the way in which individuals begin to lose their health, even if only temporarily, one must study populations prospectively. Given that many variables must be monitored simultaneously and measured repeatedly over time, a vast undertaking is required. The samples used must be large and representative of human populations at large. One way to realize an effort of this magnitude is to integrate research with public health and health care at the community level. The approaches must be applicable across cultures and must rely on procedures capable of yielding valid information across cultures. This is particularly important for social and psychological variables. Unfortunately, a major impediment to this type of effort is the notion that patients must be protected by keeping records private and confidential from those who would seek to benefit by knowledge of their plight. Obviously, privacy is one approach to the problem of unethical practices by those seeking financial gain. The downside is that barriers are created to the acquisition of knowledge. An alternative approach would be to institute severe penalties for any enterprise that misuses health status information about individuals. In other words, deal with unethical practices directly and not by creating secrecy and perpetuating ignorance.

VI. IMPLICATIONS FOR HEALTH CARE

A. HOLISTIC-EXPANDED PERSPECTIVE

Given the insights derived from psychoneuroimmunology, health care must be fundamentally a holistic endeavor, which is to say that regaining health is not simply a matter of surgical or molecular interventions. Clearly, in many circumstances, such techniques must have a role in reestablishing the balance that defines health, but exclusive reliance on such methods may prove counterproductive. This has been well documented with respect to the overuse of antibiotics and may prove to be the case for drugs used to counteract allergy or drugs aimed at psychiatric disorders, which may block symptoms while perpetuating susceptibility to the disorders. For instance, patients diagnosed with schizophrenia appear to have better outcomes in countries where pharmacological treatments are not readily available.

B. INDIVIDUALIZED HEALTH EDUCATION

The importance of fostering an understanding of the natural healing process in the patient cannot be overemphasized. This requires education of the patient in a way that is tailored to the patient's beliefs about his or her situation. How patients think about disease must be addressed in the therapeutic effort. It is also crucial to keep in mind that communication sets expectations in motion. Nocebo effects are just as powerful as those of placebos. Expectation of negative outcomes can facilitate such outcomes, so it is necessary to thoughtfully craft the messages of truth given to patients. This is complicated by the fact that in the current climate of practice, with heavy emphasis on cost containment and avoidance of liability, practitioners must increasingly approach the interaction with the patient in a brief and legalistic manner, often spending time disclosing potential adverse outcomes and obtaining signatures to the effect that the patient is in full understanding of risks. This takes away from the healing potential of the exchange by limiting the time of consultation and introducing procedures that offer nothing to the patient and are entirely designed to protect doctors, hospitals, health plans, and ultimately the pharmaceutical or medical device industries from financial liability. These priorities are misguided because patients need to be understood as individuals and treated with regard to the complex mix of influences operating in each case. This requires time and may demand a well-integrated team approach, which is not necessarily cost effective.

VII. ECONOMIC CONSIDERATIONS

Science is not the guiding force in health care, it is economics. It seems as if the health of the population was beginning to cost too much, so a whole management industry was created to control cost. However, such management complicated matters, because it could generate profit: the less spent on health care, the more profit for the health plans, the managed-care companies, and their investors. This clearly creates an unfortunate dynamic and sets the stage for a whole set of priorities that have nothing to do with the health of people.

At this juncture, the management of health care according to economic priorities presents an obstacle to a more comprehensive approach that puts the patient first and not simply treats isolated conditions such as infection, hypertension, pain, diabetes, high cholesterol, or depression exclusively by pharmacological means. Of course, clinical decision making must be made in a financially responsible way, but not simply in the interest of profit or to forestall liability.

Rectifying the dominant dynamics will require an acknowledgment at the societal level that the business model, with its natural focus on profit, is not optimal across all human endeavors. The notion that competition will bring cost down while maintaining quality in a complex human undertaking such as health care is not a well-founded expectation.

VIII. CONCLUDING COMMENTS

The scientific activity embodied in psychoneuroimmunology has demonstrated the interconnectivity of bodily systems in a way that gives credence to the notion of fundamental unity. The pathways that have been uncovered show that signaling between systems is complex and multidirectional. Any attempt to understand disease by seeking unidirectional causes that can be altered by targeting single molecules overlooks the complexity abundantly evident at all levels of analysis and the dynamic response to any intervention, with consequences that remain relatively unpredictable.

Future research must proceed along two fronts: (1) the familiar reductionistic approach to get at details and isolate mechanisms and (2) an integrative approach that tracks individuals in natural circumstances over time. The latter necessitates a coordination of epidemiology, public health, and individual health care at a global level. It will require societal changes that protect the individual without secrecy and value population health above financial gain for a minority of well-positioned individuals.

The scenario that emerges is one in which by monitoring a vast number of variables, it may become possible to isolate a handful of individuals at the highest risk of succumbing to a particular disease within a given time span, but with the realization that not all will be afflicted. Even the most extensive computation with the most powerful computers will not be able to pinpoint who will become sick before it actually happens. In effect, the individuals are acting as computing devices that cannot be surpassed by external simulations.

IX. SOURCES

Barsky, A.J. (1998), "The Paradox of Health," *New England Journal of Medicine*, 318, pp. 414–418.

Gleick, J. (1987), *Chaos: Making a New Science*, New York, Penguin Books.

Institute for the Future (2000), *Health and Healthcare 2010: The Forecast and the Challenge*, San Francisco, Jossey-Boss.

Kauffman, S. (2000), *Investigations*, New York, Oxford Press.

Keicolt-Glaser, J.K., McGuire, L., Robles, T.F., and Glaser, R. (2002), "Emotions, morbidity, and mortality: New perspectives from psychoneuroimmunology." *Annual Review of Psychology*, 53, pp. 83–105.

Larton, D., Lubahn, C., & Bellinger, D. (2001), "Introduction to biological signaling in psychoneuroimmunology," in Ader, R, Felten, D.L., & Cohen, N. (eds) *Psychoneuroimmunology*, 3rd ed, vol 1, San Diego, Academic Press.

Leff, J., Sartorius, N., Jablensky, A., Korten, A., and Ernberg, G. (1992), "The International Pilot Study of Schizophrenia: Five Year Follow-up Findings," *Psychological Medicine*, 22, pp. 131–145.

Lutgendorf, S.K. & Costanzo, E.S. (2003), "Psychoneuroimmunology and Health Psychology: An Integrative Model," *Brain, Behavior, and Immunity*, 17, pp. 225–232.

GLOSSARY

Acute phase response—The metabolic, physiologic, and behavioral changes that accompany the organism's response to infection.

Adaptation—A change at any level of organismic function that makes the organism better suited to survive within a specific context.

Adenosine—A nucleoside that is a structural component of RNA and DNA as well as being the major molecular component of energy storing molecules such as adenosine 5-triphosphate (ATP).

Adrenal gland—The small endocrine glands located above the kidneys consisting of cortex, which secretes steroid hormones (e.g., cortisol) and a medulla, which secretes catecholamines (e.g., adrenaline).

Adrenaline—An amine that functions as both a neurotransmitter in the CNS and a hormone when released from the adrenal medulla. It mobilizes the body for action.

Adrenocorticotropin hormone (ACTH) — A peptide hormone secreted by cells in the anterior lobe of the pituitary gland that stimulates the secretion of steroid hormones by the adrenal cortex.

Alexithymia—A deficit in the ability to recognize emotions in oneself as well as in others.

Allergen—A substance that produces an allergic response by causing the release of mediators of inflammation.

Allergy—The condition that renders an individual susceptible to generating an inflammatory response to innocuous substances such as pollen.

Allostatic load—A concept used to denote the level of challenge or burden that confronts the organism and demands adaptations to sustain normal function.

Amine—Any of a group of organic compounds derived by replacing hydrogen atoms in ammonia (NH_3) by organic compounds.

Amino acids—The water soluble organic acids that are composed of a carboxyl (-COOH) and an amino ($-NH_2$) group attached to the same carbon atom. There are approximately twenty such compounds, which, when linked together in short or long chains, give rise to peptides and proteins, respectively.

Amygdala—An almond shaped neural structure located within the anterior temporal lobe, which has been implicated in the regulation of emotion and the associated bodily responses.

Amyotrophic lateral sclerosis—A disease of the motor tracts and motor neurons of the spinal cord which causes a progressive atrophy of the muscles; also known as Lou Gehrig's disease.

Anaphylaxis—A massive immune response caused by a specific antigen that triggers mast cell degranulation and results in vasodilation, loss of blood pressure, and constriction of the bronchi to a degree that can be lethal.

Anergy—A decrease in the reactivity of lymphocytes to the antigen that binds to their specific receptors.

Angioedema—The swelling of deep layers of the skin, viscera mucous membranes, or within the brain.

Angiogenesis—The formation of blood vessels.

Angiotensin II—A peptide that is a potent vasopressor and causes the release of aldosterone from the adrenal cortex.

Ankylosing spondylitis (AS)—Arthritis of the spine leading to fusion of the vertebrae.

Antibody—A protein produced by B-lymphocytes in response to specific antigen.

Antidiuretic hormone—A hormone secreted by the posterior pituitary also known as arginine vasopressin that stimulates absorption of water by the kidneys.

Antigen—A segment of a protein that induces an immune response that is specific. The term refers to a substance that is an *antibody generator*. However, such substances can evoke immune responses that are not antibody mediated.

Antigen presenting cells (APCs)—A cell, such as a macrophage, that breaks up protein and displays segments on its surface so they can be seen by specific immune cell receptors and thus evoke an immune response.

Antioxidants—The substances (e.g., vitamin E, β-carotene) that inhibit oxidation reactions in the body by removing oxygen free radicals that can damage cell membranes.

Apoptosis—A process of cell death that is associated with normal development and the healthy function of tissues.

Arachidonic acid—A polyunsaturated fatty acid that acts as a precursor to certain biologically active compounds, such as prostaglandins and leukotrienes, which are involved in inflammation.

Arborviruses—Any virus that is transmitted to humans via an arthropod vector (i.e., arthropod borne virus), as when infection is acquired from a mosquito bite.

Arcuate nucleus—The small cell groups located on the ventral aspects of the hypothalamus above the median eminence and adjacent to the third ventricle.

Area postrema (AP)—One of the circumventricular structures located in the medulla oblongata.

Arginine vasopressin—See antidiuretic hormone.

Asthma—A condition characterized by narrowing of the bronchi due to inflammation and the presence of mucus in the lumen of the bronchi. It is often the result of an allergic reaction and hinders respiration.

Astrocyte—Also known as astroglia is one of the cell types that populate the central nervous system and support neurons by how they affect the neuron's microenvironment. Astrocytes get their name from their star-like appearance.

Ataxia—The loss of the ability to coordinate muscular movement.

Atopic disease—A form of hereditary allergy.

Atrial natriuretic peptide—A peptide released by cardiac atrial tissue that causes increased elimination of sodium by the kidneys.

Autism—A developmental disorder characterized by disturbances in social relatedness and language proficiency, often accompanied by stereotypic movements and mental retardation.

Autoantibody—An antibody that attacks an antigen characteristic of the organisms healthy tissues.

Autocrine—A form of self-stimulation by cells (i.e., a cell product acts back on the cell to change its activity).

Autoimmune disorder—A disorder produced by immune attack directed at the individual's own healthy tissues.

Autonomic nervous system (ANS)—The part of the peripheral nervous system that regulators visceral responses (e.g., cardiac, endocrine, gastrointestinal, circulatory responses) via the operation of two antagonistic branches—the sympathetic and the parasympathetic—causing mobilization or storage of energy, respectively.

Axon—The thread-like extension from a neuron's cell body which conducts impulses away from the cell body to other neurons or effector cells.

Axon reflex—refers to the process whereby a local change in the periphery not only initiates a response that travels into the CNS, but also causes the neural element to release substances into the vicinity. Typically, the neural element gives rise to impulses that are non-decremental (like axons) but conducted toward the cell body (like dendrites).

β-endorphin—A peptide with opioid properties released by the anterior pituitary and able to cause pain relief, regarded as an *endogenous morphine*.

Bacteria—The unicellular, prokaryotic microorganisms capable of causing disease.

Baroreceptor—A transducer of pressure caused by blood as it courses through blood vessels and the heart giving rise to a signal that travels via sensory nerves to areas of the brain that regulate blood pressure.

Basal ganglia—The term used to refer to relatively large neural masses located within the cerebral hemispheres on both sides. Major structures included are the caudate nucleus, the putamen, and the globus pallidus. They are also referred to as the *striatum* and have been implicated in the control of movement.

Basophils—A type of leukocyte that contains granules and may be a precursor to mast cells.

Bed nucleus of the stria terminalis (BNST)—The neural structure located anteriorly, which receives a projection from the amygdala and relays signals of neocortical origin to the paraventricular nucleus of the hypothalamus.

Beta-carotene—A precursor to vitamin A found in foods such as carrots and squash.

Biofeedback—A technique that allows an individual to gain some degree of control over autonomic responses, such as circulatory patterns or unconscious states, which cause muscle tension.

Bipolar disorder—A psychiatric disorder marked by periods of elation and depression of a magnitude that impairs reality testing; also known as manic-depression.

Blood-brain-barrier (BBB)—The characteristics of brain capillaries that prevent the movement of substances, particularly macromolecules, into the brain tissue.

B-lymphocytes—A type of leukocyte that when stimulated by antigen differentiates into plasma cells that produce antibodies specific to the stimulating antigen and capable of neutralizing its effects.

Bone marrow—The soft, fatty, vascular tissue filling the cavities of bones and serving as the source of white and red blood cells.

Borna disease virus—A pathogen capable of infecting brain tissue that has been associated with psychiatric disorders such as schizophrenia and bipolar illness.

Borrelia Burgdorferi—A parasite that causes Lyme disease in humans.

Brainstem—The portion of the brain that connects higher order neural structures such as the cerebral hemispheres to the spinal cord. It encompasses the midbrain and the medulla oblongata, where life-sustaining reflexes such as respiration are regulated.

Bronchial associated lymphoid tissue (BALT)—The lymphoid tissue found in the mucosa of the bronchi.

Burkitt's lymphoma—A malignant lymphoma that has been linked to Epstein-Barr virus infection.

Calcitonin—A hormone released by cells located in the thyroid gland that acts to lower blood levels of calcium.

Calcitonin gene related peptide (CGRP)—A neuropeptide found in the peripheral nervous system that promotes inflammation.

Cancer—A disorder of cell proliferation that causes the destruction of surrounding healthy tissue. It results from mutations in genes regulating cell division and cell death.

Candida albicans—A yeastlike fungus capable of infecting skin and mucosal membranes.

Capsid—The protein coat of a virus.

Catatonia—A condition of abnormal rigidity or flexibility of the musculature that is often associated with psychosis and causes people to remain in frozen postures for long intervals.

Catecholamines—The amines such as dopamine, adrenaline, and noradrenaline, which function as neurotransmitters.

Cell-adhesion molecules—The protein structures on cell surfaces that allow cells to attach and interact in close proximity.

Cell-mediated immunity—The immune responses that are directly mediated by lymphocytes.

Central nervous system (CNS)—The portion of the nervous system in vertebrates that resides within the skull and the spinal column, namely the brain and the spinal cord.

Central tolerance—The development of tolerance by deletion of self-reactive T-cells in the thymus.

Centromere—The region of a chromosome that appears to be the most constricted during cell division. Despite the name, it is not always located at the center of a chromosome; it is the site where the spindle attaches during cell division to pull a full complement of chromosomes into each daughter cell.

Cerebellum—The relatively large structure with three lobes located behind the medulla oblongata and under the cerebral hemispheres. It regulates complex movements and maintains posture and balance.

Cachectin (TNF-α)—The earlier name for tumor necrosis factor-α, a protein produced by immune cells that can induce a wasting syndrome often seen in patients with advanced cancers or AIDS.

Chaos—This refers to the behavior of systems, which, despite having a finite number of controlling variables, are not predictable from one moment to the next due to the complex and non-linear way in which the controlling variables interact.

Chemoattractant cytokines (chemokines)—The substances that cause immune cells to move in the direction of their increasing concentration.

Chemotaxis—The movement of a cell in a direction that is determined by the concentration gradient of one or more chemicals.

Chlamydia—A sexually transmitted disease caused the bacterium *chlamydia trichomatis*.

Chorea—The irregular involuntary movements of the limbs and facial muscles with a spasmodic quality.

Choroid plexus—A membrane rich in blood vessels that lines the ventricles in the brain and regulates the exchange of substances between the blood and the cerebrospinal fluid.

Chromosomes—The string like structures found in the cell nucleus and composed of DNA, protein, and RNA. They contain the genes responsible for passing on the species characteristics. When the cell is not dividing, individual chromosomes are not visible. When the cell begins division, they contract into paired chromatid strands, separate, reproduce themselves, and thus give rise to two identical cells.

Chronic fatigue syndrome—A disorder characterized by lack of energy and sleep disturbance thought to have an infectious etiology.

Circadian rhythm—The regular cycle of some biologic function with a period of 24-hours.

Circumventricular structures—Any of several regions in the brain where capillaries lack the usual characteristics that form the blood-brain-barrier and thus allow for the passage of macromolecular species.

Cirrhosis—A disease characterized by replacement of normal liver with fibrous tissue, which diminishes liver function.

Clonal selection theory—The theory that proposed that lymphocytes undergo genetic rearrangement, which gives rise to specific antigen receptors that must encounter specific antigen before they can become activated to mount immune responses.

Clone—A group of cells descended from a single parent cell that are therefore genetically identical.

Cluster of differentiation (CD)—The term used to label glycoproteins on the membrane of lymphocytes (e.g., CD4 or CD8), which serve as markers of the cells' stage of differentiation or functional characteristics.

Cohort—A defined group of subjects followed prospectively over time.

Colony stimulating factors—A group of cytokines that stimulate the production of leukocytes by the bone marrow.

Coma—A sustained state of profound unresponsiveness to external stimulation.

Complement—A group of plasma proteins that become activated in responses to the presence of pathogenic microorganisms and promote phagocytosis and cell wall perforation.

Complexity—The characteristic of a system whose state is influenced by a vast number of variables interacting in ways that are non-linear.

Concanavalin-A (Con A)—A glyco-protein extracted from the jack bean that promotes proliferation of T-lympho-cytes.

Consciousness—The state of being aware of one's environment and one's own mental activity.

Constant segment (C)—The relatively invariant part of the antibody molecule.

Corticosteroid binding globulin—A plasma protein that binds cortisol and can thereby regulate the actions of corti-sol.

Corticotropin—A hormone released by the anterior pituitary that stimulates the release of cortisol and other corticoster-oids from the adrenal cortex. It is also known as adrenocorticotropic hormone (ACTH).

Corticotropin releasing hormone (CRH)—A hormone produced in the hypothalamus that causes the release of corticotropin from the anterior pituitary cells known as corticotrophs.

C-reactive protein—A protein found in serum, which is elevated in dis-orders characterized by inflammatory processes.

Creutzfeldt-Jakob disease—A rare fatal disease of the brain characterized by pro-gressive dementia and loss of muscle control.

Crohn's disease—An inflammatory disease of the ileum.

Culture—The beliefs, values, and institu-tions that provide the framework within which individuals conduct their lives.

Cytokines—The proteins that act as inter-cellular signals in the generation and modulation of immune response as well as with respect to other functions within

neural and endocrine tissues (see Table 3.1).

Cytoplasm—The intracellular material outside of the nucleus which serves as a matrix containing the various organ-elles.

Delayed-type hypersensitivity (DTH)—A localized immune response in which antigen is presented to T-lymphocytes causing the release of cytokines that attract phagocytes and promote inflam-mation.

Delirium—A temporary state of mental confusion characterized by anxiety, dis-orientation, and hallucinations resulting from high fever, shock, or toxic expos-ure.

Dementia—The loss of intellectual facul-ties as a result of brain pathology marked by memory impairment, poor concen-tration, changes in emotional reactivity and personality, ultimately manifesting in maladaptive behavior.

Dementia praecox—An earlier term used to denote schizophrenia. See schizo-phrenia.

Dendrite—A branched extension of neurons that receives the bulk of inputs from other neurons and conducts im-pulses toward the cell body.

Dendritic cells—The antigen presenting cells, which process proteins into frag-ments and display them on their surfaces. They are found throughout the body. In the skin, they are known as Langerhan's cells or as veiled cells when moving to-wards lymph nodes.

Deoxyribonucleic acid (DNA)—A major constituent of chromosomes composed of a chain of nucleotides in which a sugar molecule, deoxyribose, and one of four bases (adenine, cytosine, guanine, or thymine) form the rungs in what appears as a spiral staircase and serves as the genetic code.

DiGeorge syndrome—A condition in which the thymus gland fails to develop.

Diversity segment (D)—A region of DNA that adds additional variability to the heavy chain of antibody molecules.

Dopamine—A catecholamine that functions as a neurotransmitter in the brain. It is also a precursor in the synthesis of noradrenaline and adrenaline.

Dorsal motor nucleus of the vagus—A cluster of neuron cell bodies located in the medulla oblongata at the level of the fourth ventricle, which gives rise to parasympathetic fibers of the vagus, the tenth cranial nerve.

Dorsal root ganglion (DRG)—The cluster of neuron cell bodies that give rise to spinal sensory nerves and are located just outside the spinal column and towards the back (i.e., dorsally).

Double-blind—A procedure used in research to keep both the subject and the investigator unaware of whether any given subject is receiving active treatment or a placebo.

Eczema—An inflammation of the skin characterized by redness, itching, and lesions that become encrusted and scaly.

Edema—An excessive accumulation of watery fluid in tissues.

Eicosanoid—Any of a group of fatty acids having 20 carbon atoms and one or more double bonds (unsaturated). They include arachidonic acid and its derivatives—the prostaglandins leukotrienes, and thromboxanes.

Electroencephalogram (EEG)—A graphic record of brain electrical activity recorded from scalp electrodes.

Encephalitis—An inflammation occurring within the brain.

Encephalopathy—Any of various diseases causing disruption or damage to the brain.

Endemic—A disease that remains stable in an area or population.

Endocrine system—The ductless glands that secrete hormones directly into blood stream and serve to regulate the function of cells at a distance. They include the pituitary, the adrenals, the thyroid, the gonads, and others.

Endogenous opioids—Any of a group of substances produced by the brain, endocrine glands, or immune cells that have effects similar to those of morphine.

Endothelial cells—The cells that form a single layer lining the inner surfaces of blood and lymph vessels as well as the heart.

Endothelial leukocyte adhesion molecule-1 (ELAM-1)—A cell adhesion molecule that allows leukocytes to attach to endothelial cells.

Enkephalins—The peptides that have opiate qualities found in nervous, endocrine, and immune tissues.

Enteric nervous system (ENS)—A network of nerve cells contained within the walls of the gastrointestinal tract and serving to regulate its function.

Enzyme—A protein that acts as a catalyst and facilitates a biochemical reaction's rate within cells of the organism.

Enzyme-linked immunosorbent assay (ELISA)—A technique for determining the amount of a protein by having an antibody to the protein bound to a surface which is then exposed to a sample. The protein of interest binds to the antibody. A second antibody that is able to bind to the protein and give off a color response is then introduced, and the color change serves as a measure of concentration.

Eosinophil—A type of leukocyte containing granules capable of destroying parasites and often found elevated in the blood of allergic individuals.

Epidemiology—The field of study that examines the distribution of diseases across populations and contexts in the interest of identifying conditions or factors that increase the probability of disease.

Epinephrine—See adrenaline.

Epitope—The region of an antigen capable of combining with antigen receptors on lymphocytes and activating them to mount an immune response.

Essential nutrients—The molecules that are required for cellular metabolism but are not synthesized within the body and must be obtained from dietary sources. They include certain amino acids, fatty acids, as well as vitamins.

Eukaryote—A single cell or multicellular microorganism whose cells contain a distinct membrane bound nucleus.

Exocytosis—The movement of vesicle enclosed molecules from inside the cell to the extracellular fluid by fusion of the vesicular membrane to the cell's membrane and discharge of the contents into the cells microenvironment.

Extravasation—The passage of substances and cells from vessels into the surrounding tissue.

Extraversion—A dimension of personality primarily characterized by an outgoing social style.

Fab region—The part of an antibody molecule that contains both heavy and light chains.

Fatty acid—An organic compound consisting of a hydrocarbon chain and a terminal carboxyl group (-COOH), with chain length ranging from a single carbon to nearly 30 carbon atoms. Large chain fatty acids may be saturated (no double bonds) or polyunsaturated (two or more double bonds).

Fc region—The part of an antibody molecule that only contains constant segments of the heavy chain.

Fibrinogen—A protein in blood plasma that is essential for the coagulation of blood.

Fibroblasts—The cells that are present in connective tissues and secrete collagen, an insoluble fibrous protein.

Fibromyalgia (FM)—A syndrome characterized by chronic pain in the muscles, tenderness at specific sites in the body, and fatigue.

Flatworms—Any of a variety of worms, including parasitic ones such as tapeworms and flukes, which have a soft flat, bilaterally symmetrical body. They belong to the phylum Platyhelminthes.

Flora—The microorganisms that normally inhabit the gastrointestinal tract and help maintain normal function.

Flow cytometry—A technique for enumerating cell types by measuring individual cell characteristics as they flow in single file through the device.

Follicular stimulating hormone (FSH) —A glycoprotein released by the anterior pituitary that acts on the ovaries to assist in follicle maturation and the secretion of estradiol. In males, it induces spermatogenesis.

Free-radical scavenger—A molecule that acts to neutralize free-radicals, normal by-products of oxidation reactions in metabolism.

Fungus—A group of eukaryotic organism once regarded as plants but now having their own separate kingdom, *Fungi*. They exist as single-celled or multicellular organisms that can be parasitic.

Galanin (GAL)—A peptide found in nerve terminals, especially within the peripheral nervous system.

Gamma aminobutyric acid (GABA)—An amino acid found in the central nervous system that functions as an inhibitory neurotransmitter.

Gammaglobulin—The class of plasma proteins that function as antibodies.

Ganglia—The clusters of neuron cell bodies located outside of the central nervous system.

Gastritis—An inflammation of the mucosal surface of the stomach.

Gastroenteritis—An inflammation of the mucosal surfaces of the stomach and intestines.

Gene—A segment of DNA coding for the sequence of amino acids in a particular protein or closely related proteins with a specific function within the organism.

General paresis—A brain disease occurring as a late consequence of syphilis characterized by dementia and progressive muscular weakness leading to paralysis.

Genome—The complete set of genes present in an organism's chromosomes.

Genotype—The specific alleles that are contained within the DNA of an organism.

Gerstmann-Strausser-Scheinker syndrome—A form of encephalopathy leading to dementia and possibility caused by a prion.

Glial cells—The cells of the central nervous system that support the neurons. They include astroglia, oligodendroglia, microglia, and ependymal cells.

Glomerulonephritis—An inflammation of the glomeruli that is not the result of infection in the kidneys.

Glossopharyngeal nerve (IX)—The ninth cranial nerve emerging from the medulla oblongata mediating sensory and motor signals to the area of the throat.

Glucagon—A peptide hormone that stimulates breakdown of glycogen by the liver and increases the blood sugar level.

Glucocorticoid—A steroid compound like cortisol released by the adrenal cortex that acts as a hormone modulating metabolism and has antiinflammatory properties.

Glycoprotein—A protein with a covalently linked carbohydrate. They are important components of cell membranes and many function as hormone receptors.

Gonorrhea—A sexually transmitted disease caused by gonococci and affecting primarily the genitourinary tract.

gp 120—A surface glycoprotein found in the envelope surrounding the genetic material of human immunodeficiency virus (HIV).

Gram-negative—The bacteria that are not stained by the Gram violet stain.

Gram-positive—The bacteria that are stained by the Gram violet stain.

Granulocyte-macrophage colony stimulating factor (GM-CSF)—A cytokine that promotes the production of leukocytes containing granules in the cytoplasm.

Granulocytes—A group of leukocytes that have granules in the cytoplasm.

Granuloma—A mass of inflamed tissue contained within a granule formed in response to infection.

Granulopoiesis—The formation of granulocytes from stem cells in the bone marrow.

Grave's disease—A condition produced by excessive thyroid hormone and characterized by protrusion of the eyeballs and anxious restlessness.

Growth hormone (GH)—A peptide hormone released by the anterior pituitary that influences metabolism and growth by promoting the release of insulin like growth factor-1 (IGF-1).

Growth hormone releasing hormone (GHRH)—A peptide released from the hypothalamus that promotes the secretion of growth hormone from the anterior pituitary.

Guillain-Barré Syndrome—A neurologic syndrome usually presenting after a virus infection marked by tingling sensations in the limbs muscular weakness and a flaccid paralysis.

Gut associated lymphoid tissue (GALT)—The lymphoid tissue located in the mucosa of the gastrointestinal tract.

Hapten—A molecule that can be immunogenic by binding to an endogenous protein, giving it the appearance of foreign antigen, but that is otherwise inactive.

Hardiness—A construct used to describe individuals who believe they have the power to overcome challenges, see them as opportunities for personal growth, and see life as meaningful.

Hashimoto's thyroiditis—An autoimmune disease of the thyroid gland causing hypothyroidism.

Health—The overall condition of an organism, which at any given time denotes freedom from disease or abnormality.

Heavy-chain—The longer of the chains composing the antibody molecule.

Heavy-chain class switching—The process whereby the constant region of the heavy chain changes from an IgM isotype to either IgA, IgG or IgE in response to antigen.

Helminthes—Any worms that are parasitic, such as roundworms and tapeworms.

Hematogenous—Originating in or spreading via the blood.

Hematopoiesis—The formation of blood cells principally occurring in the bone marrow.

Hematopoietic stem cells—The cells that become differentiated into the various cell types in the blood.

Heparin—A complex organic acid found in the lung and liver that prevents platelet agglutination and blood clotting.

Herd immunity—The protection of a population to a given infection that occurs when susceptible individuals are dispersed among those who are resistant thereby hindering transmission.

Herpes virus—One of several DNA-containing viruses causing a wide range of latent infections such as genital herpes, shingles, and life-threatening conditions such as encephalitis.

High endothelial venule cells (HEV)—The cells composing the inner wall of venules in lymphoid tissue through which lymphocytes migrate into the lymphoid tissues.

Hippocampus—A neural structure embedded within the temporal lobes of the brain that has been included as a component of the limbic system and plays a prominent role in long-term-memory formation.

Histamine—An amine found in the nervous system and mast cells serving to mediate inflammatory responses and allergic reactions. It is a potent stimulant of gastric secretions, bronchial constriction, and vasodilation.

HIV (human immunodeficiency virus)—An RNA containing virus (retrovirus) that has a particular affinity for T-lymphocytes and is the cause of AIDS.

Hives—A skin condition characterized by welts that itch intensely and can be due to allergy, infection, or an emotional state.

Holism—The perspective that views the organism as an integrated structure that cannot be treated as simply an aggregate of independent systems.

Homeopathy—A system of treating disease by using minute, almost vanishing doses of substances, which, in large amounts, produce symptoms in healthy persons similar to those of the disease.

Hormone—A substance, often a peptide or a steroid produced by cells within the CNS or endocrine glands and released into the blood stream, which causes systemic effects in metabolism and physiological activity.

Human chorionic gonadotropin hormone (hCG)—A glycoprotein produced by the placenta that acts to stimulate ovarian release of estradiol and progesterone during pregnancy.

Human leukocyte antigen (HLA) system—The two groups (Class I and II) of surface glycoproteins that function as markers of self at the cellular level and participate in antigen presentation to immune cells.

Humoral immunity—The aspect of the immune response that relies on the production of specific antibody by activated B-lymphocytes known as plasma cells.

Hyperphagia—A disorder marked by excessive and incessant food consumption.

Hypersomnia—A disorder marked by excessive sleep.

Hyperthyroidism—A disorder marked by excessive production of thyroid hormone, leading to increased basal metabolism.

Hypnosis—A process that induces a trance-like state in which the subject may be highly suggestible and experience hallucinations or behave in ways that are externally commanded.

Hypoglycemia—A low concentration of glucose in the blood.

Hypophysectomy—The removal or destruction of the hypophysis, also known as the pituitary gland.

Hypophysis—The organ, also known as the pituitary gland, which is located at the base of the brain and is the master gland of the endocrine system. It stimulates the release of hormones, which promote growth and regulate metabolic activities.

Hypothalamic-pituitary-adrenal axis (HPA axis)—The system primarily concerned with the regulation of the release of adrenocortical steroid hormones.

Hypothalamus—A relatively small region of the brain located at its base, directly under the thalamus and serving as a center for coordinating most life sustaining responses including cardiovascular, thermal, endocrine, autonomic, ingestive, and sexual activities.

Hypothyroidism—The insufficient production of the thyroid hormone.

Idiotype—The structure of the variable region of an immunoglobulin, which gives it the ability to bind specific antigens.

Idiotyte/anti-idiotype interactions—The fact that the possibility exists for antibodies to see the idiotype of other antibodies as antigen and bind it in ways that can block or stimulate the bound idiotype.

Immune complex—A combination of an antibody and antigen or antibody-antigen and complement, which can accumulate on cell surfaces and tissues causing a chronic state of inflammation such as vasculitis, endocarditis, neuritis, or glomerulonephritis.

Immunodeficiency—A condition of immune unresponsiveness characterized by increased susceptibility to infection.

Immunoglobulins (Ig)—The large glycoproteins secreted by plasma cells that function as antibodies to specific antigens thereby neutralizing pathogens or their toxins.

Immunoglobulin-supergene family—The genetic loci that code for cell adhesion molecules on endothelial cells such as intercellular cell adhesion molecule-1 (ICAM-1), which allow passing leukocytes to become attached and move into the extravascular space.

Inflammatory process—A localized protective reaction to tissue irritation, injury, or infection involving leukocytes that causes pain, redness, swelling, and may impair tissue function.

Insular cortex—An area of neocortex located laterally and hidden by the temporal lobes.

Insulin—A polypeptide hormone secreted by the islet by Langerhans in the pancreas, which helps regulate blood glucose and its storage as glycogen as well as the storage of fats as neutral lipids and further promotes protein synthesis.

Insulin-like growth factor-1 (IGF-1)—A hormone released by the liver in response to growth hormone which regulates growth and development

Integrins—The cell surface proteins that promote adhesion.

Intercellular adhesion molecule-1 (ICAM-1)—A cell surface glycoprotein that acts as a receptor for selectins and integrins expressed on leukocytes to promote adhesion.

Interdigitating cell—A dendritic cell occurring in the thymus, lymph nodes, or the spleen whose primary function is antigen presentation.

Interferons (INF)—The glycoproteins produced by a variety of cells including leukocytes that block viral replication in infected cells and act as cytokines to modulate immune responses (see Table 3.1).

Interleukins (ILs)—A growing group of cytokines that have been implicated in the regulation of all facets of immune responsiveness, but which are also found in other tissues such as the brain (see Table 3.1).

Intermediolateral column—The cell column that gives rise to the lateral horn of the grey matter of the spinal cord extending from the first thoracic to the second lumbar segment, containing the preganglionic autonomic neurons of the sympathetic nervous system.

Intracellular messengers—The molecules that respond to membrane receptors and initiate activity within the cell that can modify genetic expression and thereby alter the cell's function.

Introversion—A dimension of personality primarily characterized by inhibition particularly in social situations.

Ionizing radiation—Any radiation of sufficiently high energy to cause ionization (i.e., the release of electrons from water molecules), thus creating a highly reactive species capable of tissue damage.

Islet of Langerhans—The irregular clusters of endocrine cells scattered throughout the pancreas that secrete insulin and glucagon.

Joining segment (J)—Gene segments that contribute to diversity of the light and heavy chains of antibodies.

Kinins—A group of peptides involved in inflammation produced by the splitting of blood plasma globulins, such as the kininogens.

Kuru—A progressive encephalopathy that is fatal and is probably caused by a prion.

Latent viruses—Any viruses that can persists in the body without causing clinical symptoms but can become reactivated particularly in response to stress.

Leukocytes—The cells in the blood that are components of the immune system. They include neutrophils, lymphocytes, monocytes, eosinophils, and basophils. Also referred to as white blood cells.

Leukocytosis—An abnormally large increase in the number of leukocytes (white blood cells) in the circulation usually a sign of infection, inflammation, or some other form of stress.

Leukotrienes—The products of arachidonic acid metabolism that function as mediators of inflammation and participate in allergic responses.

Light-chain—The shorter chain of antibody molecules that contains a variable and a constant region.

Limbic system—A group of bilateral neural structures including the amygdala, hippocampus, nuclei in the hypothalamus, and thalamus as well as cortical structures such as the cingulate gyrus, which have been implicated in the regulation of emotion and associated autonomic responses.

Lipids—The organic compounds that are not soluble in water. They serve to store energy and are major constituents of cell membranes. They also serve as precursors to steroid hormones.

Lipocortin-1—A protein produced by leukocytes in response to cortisol that prevents the entry of leukocytes into sites of inflammation.

Lipoprotein—The compounds consisting of a lipid combined with a protein. They are the main structural component of cell membranes and function to transport lipids as well. Blood cholesterol is regulated in part by low-density (LDL) or high-density (HDL) lipoproteins. The latter is most effective in transporting cholesterol to the liver.

Luteinizing hormone—A hormone produced by the anterior pituitary that stimulates the production of testosterone from the testes in males and ovulation in females.

Lyme disease—An inflammatory disease caused by a spirochete transmitted by ticks characterized initially by a rash and flu like symptoms that can give way to arthritis, cardiac dysfunction, and psychiatric disturbances.

Lymph nodes—The small, oval shaped bodies located along the lymphatic vessels, which are sites wherein lymphocytes encounter antigen and become activated to mount immune responses.

Lymphocyte—A type of leukocyte that includes B-cells, T-cells, and NK cells, which are central to humoral, cell-mediated, and innate immune responses, respectively.

Lymphocyte function associated antigen-1 (LAF-1)—A cell surface integrin that allows the attachment of T-cells to other cells.

Lymphoid follicle—A spherical mass of lymphoid cells found in the lymph nodes.

Lymphotoxin (TNF-β)—An earlier name for tumor necrosis factor-β given due to its toxic effect to certain susceptible cells.

Macrophages—The large phagocytes that are tissue bound and function as antigen presenting cells.

Major histocompatability complex (MHC)—The name given to the genes coding for the cell surface histocompatability antigens in non-human organisms that are analogous to the HLA system in humans.

Malaria—An infectious disease caused by parasitic infection of red blood cells by a protozoan of the genus *Plasmodium*, which is transmitted by the bite of an infected female anopheles mosquito.

Mammotrophs—The cells in the anterior pituitary that release prolactin.

Mast cell—A type of immune cell found throughout the body, which releases proinflammatory substances such as histamine in response to irritants including allergens.

Median eminence—A circumventricular organ located at the base of the hypothalamus, where the blood brain barrier is porous.

Megakaryocyte—A large cell of bone marrow origin which gives rise to blood platelets.

α-Melanocyte stimulating hormone (α-MSH)—The peptide hormone secreted by the anterior pituitary that stimulates melanin synthesis in melanocytes and influences skin color.

Melanoma—A malignant tumor arising from melanocytes in the skin and capable of widespread metastasis.

Melatonin—A hormone produced by the pineal gland.

Meningitis—An inflammation of the meninges membrane surrounding the brain and spinal cord.

Menopause—The period marked by the natural cessation of fertility in females occurring in late middle age.

Mental retardation—The subnormal intellectual function that renders an individual unable to function cognitively at or above 98% of the population within their age group (i.e., individuals with IQs of 69 or below).

Mesenteric/celiac ganglia—These are prevertebral clusters of postganglionic cell bodies located distally from the spinal chord conveying sympathetic nervous system signals to the gastrointestinal and genitourinary tracts.

Meta-analysis—A method of statistical analyses used to evaluate whether the findings across a wide range of studies consistently demonstrate that a given effect is real. For instance, published studies consistently show that depressed individuals have a higher white blood cell count.

Metastasis—The migration of cancerous cells from the original site of malignant transformation to one or more sites elsewhere in the body usually by way of blood vessels or the lymphatic vessels.

Methylation—The attachment of methyl groups to DNA at points where cytosine and guanine are adjacent which acts to block transcription.

Microglia—The small cells in the central nervous system of bone marrow origin serving as antigen presenting cells and promoters of inflammation at sites of neural damage.

Migration inhibition factor—A cytokine that inhibits the movement of leukocytes out of an area of infection.

Mineral—An inorganic element, such as calcium, iron, potassium, sodium, or zinc that is essential to cell function.

Mineralocorticoids—The steroid hormones secreted by the adrenal cortex that regulate the balance of water and electrolytes in the body.

Mitogens—Agents that non-specifically stimulates lymphocyte proliferation.

Molecular mimicry—The fact that antibodies generated to a foreign antigen can cross react with healthy self tissue due to the similarity in molecular structure between the foreign and self antigen.

Monoaminergic neurons—The neurons that contain one of the amine neurotransmitters such as dopamine, noradrenalin, or serotonin.

Monocytes—A large circulating phagocyte with a single well-defined nucleus that may be a precursor to the tissue bound macrophage.

Mononuclear leukocytes—The subpopulation of leukocytes with a single well defined nucleus, specifically monocytes, and lymphocytes.

Multiple sclerosis—An autoimmune disorder in which myelin is destroyed in the central nervous system, causing muscle weakness, incoordination, as well as disturbances in vision and speech.

Mutation—A change in the genetic material of a cell so that its function or structure may be altered in a way that is positive or potentially detrimental. Mutations may be restricted to an individual or be transmitted across generations if they occur within the germline cells.

Myasthenia gravis—An autoimmune disorder in which the acetylcholine receptor is blocked causing generalized weakness of the muscles and fatigue.

Mycosis—A fungal infection in or on a part of the body, which causes disease.

Myelin—A phospholipid produced by oligodendroglia in the CNS or Schwann cells in the periphery, which insulates axons and enhances the conductivity of impulses.

Natural killer (NK) cells—The granular lymphocytes capable of cytotoxicity directed at virally infected or malignant cells in a non-specific manner.

Naturopathy—A system of healing that avoids surgical or pharmacological interventions and emphasizes only naturally occurring remedies.

Neocortex—The more recently evolved regions of the cerebral cortex that greatly expand the number of neurons in the brain arranged within a six-layered structure on the surface of the brain.

Nephelometry—A technique that relies on light defraction to measure the size and concentration of particles in a fluid.

Nerve—The chord-like bundle of nervous tissue made up of myelinated and unmyelinated fibers held together by connective tissue carrying impulses to and from the central nervous system.

Nerve growth factor (NGF)—A protein that stimulates the growth of neurons.

Neurasthenia—A complex of symptoms including chronic fatigue, weakness, loss of memory, and generalized arches and pains.

Neuritic plaque—A spherical mass of amyloid fibrils surrounded by distorted interwoven neuronal processes found in the brains of persons with dementia (e.g., Alzheimer's disease).

Neurofibrillar tangles—The disruption of the normal organization of long, thin microscopic fibrils that run through the extent of neurons and give the neuron structural support and shape.

Neuroimaging—The techniques that permit the visualization of brain structure or its activity in a non-invasive manner by monitoring emissions of radioactive tracer substances or normal constituents of brain tissue energized by external energy fields. Mathematical methods allow three-dimensional reconstructions of the detected activity by sensors outside the skull.

Neurokinins—The neuropeptides of the tachykinin family, which also includes substance P. They are contained in sensory neurons and mediate neurogenic inflammation.

Neuroleptic—A class of drugs normally used to treat psychotic disorders.

Neuron—A cell of the nervous system specialized for conducting impulses, which is characterized by a dendritic tree, a cell body, and an axon. Conduction proceeds from dendrites, where inputs are summated to the initial segment of the axon (axon hillock), where a non-decremental impulse is generated.

Neuropeptide—Any of various peptides found in neural tissue often colocalized with classical neurotransmitters.

Neuropeptide Y (NPY)—A neuromodulator peptide found within the hypothalamus and colocalized with catecholamines in the sympathetic nervous system.

Neurophysin—Any of a group of proteins synthesized in the hypothalamus that transport and store hormones.

Neurosyphilis—A nervous system manifestation of syphilis.

Neuroticism—A dimension of personality characterized by a propensity to experience negative emotions.

Neurotoxins—Any chemical that either physically damages a neuron or reduces its ability to function normally.

Neurotransmitter—A chemical that transmits neural activity across synapses.

It is contained in vesicles in the pre-synaptic terminal, released by the arrival of action potentials, and after diffusing across the synaptic cleft binds to postsynaptic receptors thereby causing a transient change in membrane potential due to the flow of ions in or out of the postsynaptic neuron.

Neutrophil—A type of leukocyte that is abundant in the circulation and acts as a phagocyte throughout the body.

Nitric oxide (NO)—A gaseous modulator of tissue activity including dilation of blood vessels, neurotransmission, and immune effector activity against tumor cells and pathogens.

N-methyl-D-aspartate (NMDA)—A glutamate receptor that plays a critical role in experience dependent plasticity.

Nocebo effect—The opposite of placebo, that is, a disturbance caused by the expectation that some compound, activity, or event causes damage.

Norepinephrine—Also known as nor-adrenaline, is an amine released by both the adrenal medulla and neurons in the central nervous system and the sympathetic nervous system, which induces arousal.

Nucleic acid—A complex organic compound consisting of a chain of nucleotides giving rise to two main types deoxyribonucleic acid (DNA) and ribonucleic acid (RNA).

Nucleotide—A compound consisting of a sugar (ribose or deoxyribose), purine, or pyrimidine base combined with a phosphate group and forming the basic units of DNA and RNA.

Nucleus—(1) A group of nerve cells with a distinctive organization found within the central nervous system; (2) A membrane bound structure located within a cell and containing the chromosomes carrying the genetic material.

Nucleus basilis—A neural structure located in the diencephalon that is a major source of cholinergic neurons innervating the neocortex.

Nucleus of the solitary tract (NTS)—A neural structure in the medulla oblongata that relays visceral afferent signals to higher order neural structures in the brain.

Oligodendroglia—Also known as oligodendrocyte, is a form of CNS accessory cell associated with formation of the myelin sheath that insulates axons and speeds up their conduction of action potentials.

Omega-3 (ω-3) fatty acid—A fatty acid with a double-bond on the third carbon atom found in fish oils that is protective with respect to cardiovascular disease.

Omega-6 (ω-6) fatty acid—A fatty acid with a double-bond on the sixth carbon atom which is much less protective with respect to cardiovascular disease and inflammatory disorders in general.

Oncogene—A gene that when mutated promotes the malignant transformation of cells.

Orbito-frontal cortex—The inferior portion of the frontal lobe of the brain lying just above the eye sockets.

Organum vasculosum of the lamina terminalis (OVLT)—A circumventricular structure located frontally in the cerebrum where the blood brain barrier is porous.

Osmoreceptor—The neural structures in the CNS, which respond to changes in osmotic pressure generated by concentration differences between extra and intracellular fluids.

Osteopathy—A system of healing based on the theory that disturbances in the musculoskeletal system affect the function of other organs and cause disorders that can be corrected by manipulative techniques in conjunction with conventional medicine therapeutics.

Oxidative metabolism—The reactions that involve the transfer of electrons from some compound to oxygen molecules.

Oxygen free radical—A highly reactive species of oxygen due to the occurrence of unpaired electrons, a normal byproduct of oxidation reactions in metabolism.

Oxytocin—A peptide hormone released from the posterior pituitary that promotes contraction of the uterus during labor, facilitates the release of breast milk, and may foster attachment.

p 53 gene—The gene coding for protein that protects cells from the effect of DNA damage. It is commonly deleted or mutated in cancer cells.

Parabrachial nuclei—Neural structures located in the brainstem that project to the amygdala and appear involved in the regulation of autonomic responses.

Paracrine—The local effects on nearby cells of substances released by other cells.

Paranoia—A psychiatric condition characterized by delusions of persecution or an extreme distrust of others.

Parathyroid hormone (PTH)—A peptide hormone release by the parathyroid glands that regulates the level of calcium in the blood.

Paraventricular nucleus (PVN)—A triangularly shaped nucleus located in the anterior hypothalamus, which is functionally implicated in endocrine and autonomic regulation.

Parkinson's disease—A progressive nervous system disease in which dopamine neurons are destroyed and which is characterized by muscular tremor, difficulty initiating voluntary acts, and partial facial paralysis.

Pediatric autoimmune neuropsychiatric disorders (PANDAS)—The disturbances in behavior that appear to be triggered after infection and may reflect cross reactivity between pathogen antigens and self-antigens normally occurring in neural tissue.

Peptide—Any compound consisting of two or more amino acids linked by bonds formed by the carboxyl (-COOH) group of one amino acid with the amino (-NH$_2$) group of the other. Polypeptides contain more than 10 amino acids and usually contain between 100 and 300.

Periaqueductal grey—The area surrounding the cerebral aqueduct that has been implicated in the processing of pain stimuli.

Periarteriolar lymphocyte sheath (PALS)—The aggregation of lymphocytes around the arterioles within the spleen.

Peripheral nervous system (PNS)—The nervous tissue outside of the bony structures that contain the central nervous system. It includes cranial nerves, spinal nerves, the autonomic nervous system and the enteric nervous system.

Peripheral tolerance—The development of tolerance for self-antigens that occurs due to anergy or suppression of self-reactive lymphocytes that have not been deleted in bone marrow or thymus.

Personality—The stable behavioral characteristics of an individual.

Pervasive developmental disorders (PDD)—The conditions, which are characterized by deficits in social development in combination with disturbances in motor functioning, language, and self-regulation of emotion and behavior.

Phenotype—The observable characteristics of an organism determined by the interactions of genes and the effects of the environment.

Phytohemagglutinin (PHA)—A plant-derived substance that promotes the proliferation of T-lymphocytes.

Pineal gland—A small glandular structure located between the superior colliculi of the midbrain which releases the hormone melatonin.

Pituitary adenylate cyclase activating polypeptide (PACAP)—One of several peptides that appear to be localized within peripheral nervous tissue and are released along with the classical neurotransmitters.

Placebo—A neutral substance or procedure that has positive effects as a result of the expectation that it is an effective treatment, particularly if both the doctor and the patient believe in the treatment's effectiveness.

Plasma cells—An antibody producing B-lymphocyte after activation by specific antigen.

Plasmapheresis—A process whereby the clear fluid portion of the blood is removed while the remaining cellular components are returned to the donor.

Platelet activating factor—A substance released by a variety of cells including leukocytes that causes the aggregation of platelets and is involved in the deposition of immune complexes.

Pleiotropy—A term used to denote that a specific molecule has more than one effect.

Plexus—A structure in the form of a network composed of nerve fibers, blood, or lymphatic vessels.

Pokeweed mitogen (PWM)—A substance that stimulates the proliferation of B-lymphocytes in a non-specific fashion.

Polymerase chain reaction (PCR)—A technique used to make copies, that is, amplify, a fragment of DNA. The technique relies on the enzyme polymerase which catalyzes the formation of polynucleotides using an existing strand of DNA as a template in a medium containing a supply of free nucleotides. Cycles of heating and cooling are used to separate the DNA strands and drive the cycles of replication.

Posterior pituitary—The region of the pituitary gland that releases the hormones arginine vasopressin and oxytocin.

Postganglionic neurons—The neurons whose cell bodies are located in prevertebral or paravertebral ganglia and project to visceral tissues.

Preganglionic neurons—The neurons whose cell bodies are located within the brainstem or spinal cord and project to ganglia where postganglionic neurons have their cell bodies.

Preoptic area—The region of the hypothalamus anterior to the optic chiasm.

Procaryotes—The organisms whose cells lack a nucleus and have DNA that is not organized into chromosomes.

Progesterone—A steroid hormone that acts to prepare the uterus for implantation of the fertilized ovum to promote pregnancy and the development of the mammary glands.

Proinflammatory cytokines—The proteins released by immune cells that promote inflammatory responses. Major ones include IL-1β, IL-6, and TNF-α.

Prolactin (PRL)—A pituitary hormone that stimulates and sustains the secretion of breast milk.

Pro-opiomelanocortin (POMC) protein—The protein that is cleaved to produce the hormones corticotropin, B-endorphin and melanocortin.

Prospective studies—The investigations in which the population of interest is studied over time beginning prior to the occurrence of some disorder or other condition. They are typically used to identify predictors or early manifestations of the disorder of interest.

Prostaglandins—The substances derived from arachidonic acid in a variety of tissues that cause the contraction of smooth muscle and participate in the inflammatory response.

Proteases—The enzymes that catalyze the hydrolytic breakdown of proteins.

Protein—The macromolecules composed of long chains of amino acids. They occur in two basic forms: (1) globular proteins, which include enzymes, antibodies, and some hormones; or (2) fibrous proteins, which include structural proteins such as collagen, myosin, and fibrin.

Protozoa—Any of a group of single celled eukaryotic microorganisms such as amoebas, ciliates, flagellates, and sporozoans.

Psychosomatic disorder—A disorder characterized by physical symptoms resulting from psychological factors involving any system of the body, such as the gastrointestinal tract, for example.

Random assignment—A method for assigning subjects to treatments whereby every member of the target population has an equal chance of being selected for any of the treatments.

Raphe nuclei—The clusters of neurons in the brain stem that give rise to the serotonergic innervation of the diencephalon, limbic structures, and neocortex.

Receptors—A molecular structure or site on the surface or interior of a cell that binds another molecule and together produce some change in cell function.

Recombinase—An enzyme that catalyzes the recombination of chromosomal material.

Reductionism—An approach to seeking scientific understanding by assuming that complicated phenomena or structures can be analyzed into a few basic constituents linked together by relatively simple rules.

Releasing hormones—The peptides originating within hypothalamic neurons that promote the release of stimulating hormones from the anterior pituitary ultimately leading to the secretion of hormones by the peripheral endocrine glands.

Reticular formation—An extensive network of neurons centrally located particularly within the brain stem, whose activity serves to regulate cortical arousal.

Retrospective studies—Investigations conducted by asking a subject or otherwise obtaining information about past events in an effort to gain understanding about what may have been a factor leading to the subject's present condition or status.

Retroviruses—A family of viruses that contain RNA and reverse transcriptase (e.g., HIV).

Reye's syndrome—An acute encephalopathy characterized by fever, vomiting, disorientation, and coma occurring typically in children following a viral illness, especially if they have been treated with aspirin.

Rheumatoid arthritis—A chronic and progressive disease characterized by stiffness and inflammation of the joints sometimes leading to deformity and disability.

Rheumatoid factor (RF)—An antibody to the Fc portion of immunoglobulins that is often found in the serum of individuals with inflammatory disorders.

Ribonucleic acid (RNA) editing—A process of altering RNA before it is transcribed into protein, thus changing protein structure from what is coded in the DNA.

Rickettsias—A type of bacteria carried as parasites by ticks, fleas, and lice that can cause diseases, such as typhus and Rocky Mountain spotted fever.

Roundworm—A parasitic worm of the Nematode class.

Schizophrenia—A form of psychosis characterized by delusions, hallucinations, social withdrawal, and blunted affect.

Schwann cell—Any of the cells that produce myelin to insulate axons in the peripheral nervous system.

Scleroderma—A pathological thickening and hardening of the skin caused by swelling and growth of fibrous tissue.

Selectin—A class of proteins that act as cell adhesion molecules.

Self-similarity—A concept used to convey the notion that there is symmetry across scale. It implies that there is pattern inside of pattern and that complexity is retained across levels of description.

Sensitization—The process of inducing an allergic responses by exposure to an allergen.

Sepsis—The presence of pathogenic organisms or their toxins in significant concentrations in the blood or in the tissues.

Septal area—A neural structure located medially and anterior to the thalamus that is interconnected with structures of the limbic system and may be involved in the modulation of emotional displays.

Septic shock—The collapse of vital physiological functions due to massive infection, which can be lethal.

Seropositive—The status of having antibodies in blood serum specific for a particular pathogen.

Serotonin—A monoamine neurotransmitter implicated in the regulation of emotional tone and impulse control. It is also found in other tissues such as the gastrointestinal tract and immune cells. It participates in vasoconstriction and smooth muscle activity.

Sickness behavior—A behavioral profile associated with acute infection consisting of sleepiness, fatigue, and social withdrawal.

Sjögren's syndrome—An autoimmune disorder characterized by inflammation of the cornea, decreased tear flow, and dilation of skin capillaries causing dark purple blotches on the face.

Social networks—The types of social contacts available to an individual including family, friends, organizations, and religious institutions.

Social support—The benefits derived from social contact consisting of activities such as companionship, listening, giving advice, or providing physical assistance.

Socioeconomic status (SES)—The standing or position in a social hierarchy based on occupational status, income, and educational level.

Spinal cord—The neural tissue that extends into the spinal column and is the source of the spinal nerves, which innervate the musculature and the viscera.

Spirochete—A slender, spiral, motile bacteria.

Splanchnic nerves—The nerves that project to visceral organs and convey the influence of the sympathetic nervous system.

Spleen—A large, highly vascular lymphoid organ lying to the left of the stomach and below the diaphragm. It is a major site where lymphocytes become activated by antigen arriving via the blood.

Spondyloarthropathies—An inflammation of the intervertebral articulations of the spinal column.

Stevens–Johnson syndrome—A severe inflammatory eruption of the skin and mucous membranes, usually occurring in young individuals following respira-tory infection or as an allergic reaction to drugs or other substances.

Stimulating hormones—The peptides released by cells of the anterior pituitary, which stimulate the release of hormones such as thyronine, cortisol, testosterone, or estradiol from peripheral endocrine glands.

Stress—A challenge or burden affecting an organism, which constitutes a change in its context and requires functional adjustments to sustain homeostasis.

Striatum—A set of structures within the cerebral hemispheres, also known as the basal ganglia, and consisting of the caudate nucleus, putamen, and globus pallidus, which are all implicated in the regulation of motor behavior.

Stroke—A sudden loss of some aspect of brain function due to the blockage or rupture of a blood vessel supplying a region of the brain crucial for the particular function.

Stromal cells—The cells that form the structural framework of a given organ as distinguished from the cells that perform the special function of the organ.

Subfornical organ (SFO)—A circum-verntricular structure with a porous blood brain barrier located anteriorly under the fornix.

Substance P (SP)—A short-chain polypeptide that functions as a neuromodulator especially in the transmission of pain impulses from the periphery to the central nervous system.

Substantia nigra—The bilateral pigmented neural structure in the midbrain that is a principal source of dopamine neurons projecting to the striatum, which, when damaged, gives rise to Parkinson's disease.

Super antigen—The molecular structure capable of activating a large number of lymphocytes with distinct specificities.

Suprachiasmatic nucleus—A neural structure located in the hypothalamus above the optic chiasm.

Sydenham's chorea—A condition of the nervous system usually associated with rheumatic fever and characterized by involuntary, irregular, jerky movements of the muscles of the face, neck, and limbs; also known as Saint Vitus dance.

Sympathetic nervous system (SNS)—The part of the autonomic nervous system arising from thoracic and lumbar segments of the spinal cord that acts to mobilize the organism for action in the face of threat.

Synapse—The space between an axon terminal and another neuron or effector cell to which it projects. It is the gap into which nerve impulses cause the release of neurotransmitters or neuromodulators and thereby transmit signals to other cells.

Systemic lupus erythematosus (SLE)—An autoimmune disorder affecting multiple organs and frequently causing fever, weakness, fatigue, joint inflammation, skin lesions of the face and disturbances of kidney, spleen and a variety of other organs including the brain.

Tachycardia—A rapid heart rate.

Tachykinins—A class of neuropeptides that includes substance P and participates in the neurotransmission of pain signals as well as in the facilitation of inflammation by increasing vascular permeability and smooth muscle contractions.

T-cell receptor (TCR)—The membrane bound polypeptide structure that allows T-cells to bid antigen and become activated.

Technology—Any human constructions (e.g., tools) that enhance the ability of humans to gain control over some aspects of nature.

Telomere—The end region of a chromosome which consists of repeated sequences of DNA and function to signal the completion of replication during cell division. Loss of this region of the chromosome overtime triggers cell death.

Testosterone—A steroid hormone formed in the testes and possibly by the ovary and adrenal cortex which is responsible for male characteristics.

Th 1—A subset of helper T-cells that act to promote cell-mediated immune responses.

Th 2—A subset of helper T-cells that act to promote humoral immune responses.

Thalamus—A neural structure composed of various nuclei that are involved in the relaying of sensory information to the cortex. It is located medially and above the hypothalamus.

Thymocyte—An immune cell that develops in the thymus to become a T-lymphocyte.

Thymopoietin—A peptide released by the thymic epithelium involved in the maturation of thymocytes.

Thymosin—A hormone secreted by the thymus that stimulates the development of T-lymphocytes.

Thymulin—A peptide released by the thymic epithelium which promotes the maturation of T-cell precursors.

Thymus—A primary lymphoid organ where T-lymphocytes undergo differentiation and deletion.

Thyroid stimulating hormone (TSH)—A glycoprotein hormone secreted by the anterior pituitary that stimulates the release of thyroid hormones.

Thyrotrophs—The cells in the anterior pituitary that secrete thyroid-stimulating hormone.

Thyrotropin releasing hormone (TRHN)—A peptide hormone released by hypothalamic neurons that promotes the release of thyroid stimulating hormone from the anterior pituitary.

Thyroxine (T4)—An iodine containing hormone produced by the thyroid gland that increases the rate of cell metabolism.

T-lymphocytes—The lymphocytes that undergo final differentiation in the thymus.

Tolerance—Unresponsiveness to antigen that originates within the organisms healthy tissues.

Tourette's syndrome—A neurological disorder characterized by multiple facial and other body tics including grunts or compulsive utterance of obscenities.

Toxoplasmosis—A disease caused by infection with the parasite, *Toxoplasma gondii*. If exposure is prenatal, severe central nervous system damage occurs in the offspring.

Transforming growth factor-β (TGF-β)—A cytokine with antiinflammatory properties.

Treponema pallidum—A spirochete that causes syphilis in humans.

Trigeminal—The major sensory nerve for the face that also controls the muscles of chewing. It is also known as the fifth cranial nerve.

Tumor necrosis factors (TNF)—A protein produced by leukocytes and other cells that functions as a cytokine. It is central to the acute response to infection

and is capable of destroying some cancer cells.

Tumor suppressor genes—The genes that, when mutated, lose the ability to down regulate cell proliferation and may allow the escape of malignancy into uncontrolled proliferation.

Ultraviolet radiation—The electromagnetic energy of short wavelengths that are not visible but are capable of affecting cell function.

Urticaria—See hives.

Uveitis—An inflammation of the vascular, pigmentary middle coat of the eye comprising the choroid, ciliary body, and iris, a structure known as the uvea.

Vaccination—The process of exposing a subject to a weakened or killed pathogen that stimulates an immune response in the subject, is incapable of causing severe disease in most subjects, and serves as protection from further infection.

Variable segment (v)—A gene segment coding for part of the antigen binding site formed by the light and heavy chains of antibody molecules.

Vasoactive intestinal peptide (VIP)—The peptide hormone released by cells in the pancreas but also found in nerve terminals which modulates leukocyte migration and activity.

Vector—An organism, such as a mosquito or a tick, that carries a pathogen from one host to another.

Veiled cells—The dendritic cells in the afferent lymphatic vessels that move towards the lymph nodes.

Ventral medulla—The portion of the medulla oblongata major that is a relay station for visceral afferent signals.

Ventricle—A small cavity or chamber within an organ, such as the brain.

Viremia—The presence of virus in the blood.

Virion—A complete viral particle consisting of RNA or DNA surrounded by a protein shell and constituting the infective form of a virus.

Viroid—An infectious particle, smaller than a virus that consists solely of a strand of RNA and causes disease in plants.

Virulence—The capacity of a microorganism to cause disease.

Virus—Any of various microorganisms that often cause disease and that consist of a core of RNA or DNA material surround

by a protein coat, which cannot replicate without infecting a host cell.

Vitamin—Any of various organic substances that are essential in tiny amounts for normal tissue function and are naturally obtained from plant and animal foods.

Yoga—A set of practices based on Hindu philosophy aimed at bringing about a higher state of consciousness and selfhood by engaging in a variety of body control exercises.

Zoonotic infection—An infection that is acquired by humans from other species (e.g., rabies).

INDEX

A

Academic exams, 105, 112, 118, 124–125, 127, 130, 193, 215
Acetylcholine, 68, 74, 85, 95, 122, 147, 166, 185, 189
Acquired immune deficiency syndrome (AIDS), 110, 112, 137–138, 140–141, 142
Acute encephalitis, 184
Acute meningitis, 184
Acute phase proteins, 46, 138, 195, 216
Acute phase response, 4, 193, 222
Adapt, 229
Adaptation energy, 104, 114
Adaptiveness, 99, 104, 115
Addison, 69
Adenosine, 214
Adenosine receptors, 214
Ader, Robert, 1, 18–19, 98–99, 212
Adolescence, 201
Adoption, 171
Adrenal cortex, 62, 76
Adrenal medulla, 62, 122, 147
Adrenaline, 122
α-Adrenergic receptors, 87, 89–90
β-Adrenergic receptors, 87, 90, 148, 216
Adrenocorticotropin hormone (ACTH), 59, 63–64, 68, 73, 76, 120–121
Affective disorders, 201
Afferent lymphatics, 47
Age, 184, 218, 226, 230
Aggression, 171, 185, 189
Ãgni, 13
Agreeableness, 100
Air pollution, 145
Airway hyperreactivity, 147
Alarm reaction, 104
Alcohol, 195
Alexander, Franz, 18
Alexithymic, 163
Allergens, 143, 145, 149, 151, 209, 227
Allergic disease, 143, 146–147
Allergic rhinitis, 143

Allergy, 3, 71, 97, 114, 134, 143–144, 148–150, 228, 232
  allergens, 143–144
  anaphylaxis, 143
  angioedema, 143
  asthma, 143
  atopic dermatitis (eczema), 143
  attention-deficit-hyperactivity disorder, 146
  environmental co-factors, 145
  genetic factors, 144–145
  migraine, 146
  pathogenic mechanisms, 145–147
  prevalence, 144–145
  psychosocial stress, 148–149
  rhinitis (hay fever), 143
  Stevens-Johnson syndrome, 144
  urticaria (hives), 143
Allostatic load, 108, 134
Alternative medicine, 210
Alternative pathway, 28
Altitude, 110
Alzheimer's disease, 123, 186
American Medical Association (AMA), 212
Amino acids, 34, 218
Amygdala, 60, 94–95, 99
Amyotrophic lateral sclerosis, 186
Analgesia, 122, 209
Anaphylaxis, 143, 146
Anergy, 39, 167
Anger, 125, 139, 160–161, 171, 173, 175–176
Angioedema, 143
Angiogenesis, 166
Angiotensin II, 68, 73
Animacules, 15
Ankylosing spondylitis (AS), 168
Anopheles mosquito, 137
Anorexia, 187, 189, 191, 195, 217–218, 222
Anorexia nervosa, 148
Anoxia, 103
Anterior cingulate cortex, 94
Anterior hypothalamic area, 65, 95
Anterior pituitary, 59

Antiacid preparations, 139
Antibiotic treatment, 187
Antibiotics, 3, 112, 139, 150, 197–198, 232
Antibodies, 24, 35–37, 64, 68, 70–71, 75, 89,
    125–126, 138, 142, 159, 166, 172, 180,
    182–183, 186, 190–191, 199, 201–203, 216,
    226–227
Antibody dependent cellular cytotoxicity
    (ADCC), 34
Anticholinergic drugs, 148
Antidepressant treatment, 160, 195
Antidiuretic hormone, 62
Antigen, 31, 47, 74, 78, 86, 89, 126, 128, 145,
    164–167, 169, 174, 182–183, 186, 201, 209,
    213, 231
Antigen binding site, 31, 35
Antigen presenting cells (APCs), 30, 49, 136, 147,
    182
Antihypertensive, 190
Antiinflammatory, 69, 172, 183, 217
Antiinfective vitamins, 219
Antimicrobial agents, 91, 231
Antioxidants, 219–220
Antipyretic, 69
Anxiety, 124, 163, 171, 174, 176, 188, 195, 198
Apathy, 200
Aphasia, 186
Apoptosis, 37, 91, 97, 155
Arachidonic acid metabolism, 29, 145
Arboviruses, 186
Arcuate nucleus, 64
Area postrema (AP), 60, 88, 94
Aretaeus, 143
Arginine, 218
Arginine vasopressin (AVP), 59, 68, 72–74, 121
Arthritis, 198
Arthropods, 137
Arthropod vectors, 136
Aseptic meningitis, 184
Assertive, 161
Asthma, 143–144, 148–150, 209
Asthma nervosa, 148
Astrocytes, 197, 174, 182–183, 190, 192
Asymmetric frontal cortex activity, 100
Ataxia, 186
Atherosclerotic plaques, 138
Atopic dermatitis, 143–144, 209
Atopic disease, 143
Atrial natriuretic peptide, 68
Attention, 203

Attention-deficit-hyperactivity disorder, 146,
    197–199
Atypical depression, 195
Australopithecus, 154
Autism, 148, 196–198, 201
Autism-spectrum disorders, 197
Autoantibodies, 164, 166–168, 172, 176, 188, 202
Autocrine, 42, 63
Autoimmune disorders, 19, 71, 97, 114, 135, 145,
    154, 165–170, 172–176, 190, 195, 202, 220
Autoimmune hepatitis, 165
Autoimmunity, 3, 164, 168–170, 173–174, 176,
    186, 190, 200, 202, 227
    ankylosing spondylitis, 168
    cancer, 168
    cirrhosis, 165
    clonal selection theory, 164–165
    gender, 169–170, 174–175
    genetic factors, 169
    glomerulonephritis, 165
    Grave's disease, 165, 170
    Hashimoto's thyroditis, 165, 170
    hermolytic anemia, 165
    hepatitis, 165
    infection, 167–168
    insulin dependent diabetes, 173
    multiple sclerosis, 174
    myasthenia gravis, 165
    obsessive compulsive disorder, 198–199
    pathogenic mechanisms, 165–167
    prevalence, 165
    psychosis, 173
    psychosocial stress, 170–174
    rehumatoid arthritis, 170–172, 202
    schizophrenia, 201–202
    scleroderma, 165
    Sjogren's syndrome, 165
    systemic lupus erythematosus, 173–174
    toxic chemicals, 168
    uveitis, 165
Autonomic arousal, 189
Autonomic nervous system (ANS), 17, 60, 82,
    84–85, 88, 90, 122, 147, 221, 227
Awaiting surgery, 127
Axon reflex, 88, 147
Ayurveda, 12

B
Babylonian, 10
Bacilli, 136–137

Bacteria, 136, 139, 184, 197
Bacterial meningitis, 137, 184
Balanced diet, 221
Baroreceptors, 73
Basal ganglia, 199
Basophils, 25, 28–29, 146
B-complex vitamins, 219
Beattie, Brow and Long, 17
Bed nucleus of the stria terminalis (BNST), 94–95
Behavior, 82, 84, 180, 185, 188–192, 194, 197,
    231
Behavioral disorders, 4, 185, 188, 191, 194, 199
Behering, Emil, 16
Beliefs, 208–209, 228
Bereavement, 124, 127, 170, 173
Berczi, I., 105
Bernard, Claude, 16–17
Beta-carotene, 219–220
Big Five, 100
Biofeedback, 215
Biopsychosocial model, 18
Bipolar disorder, 191, 201, 203
Bizarre behavior, 188–189
Black bile, 155
Black Death, 134
Bleuler, Eugen, 199–200
Blood-brain barrier (BBB), 60, 88, 174, 180–182,
    190, 192
Blood pressure, 212
Blood pressure regulation, 68
B-Lymphocytes, 25, 30–31, 33, 43–44, 49–50,
    54, 65–66, 89, 125, 145–146, 166, 172, 176,
    182, 199, 216
Body temperature, 64, 88
Bone marrow, 25, 46–48, 86
Borna virus, 185, 201
Borrelia Burgdorferi, 137, 187
Bostock, 143
Botulism toxin, 185
Brain, 82, 113, 129 , 173–174, 181–183,
    185–186, 188–190, 192, 197, 199, 203,
    207–208, 211, 214, 221, 227, 230–231
Brain abscesses, 185
Brainstem, 60, 73, 78, 93, 193
Brain trauma, 96
Brain tumor, 96
Breast cancer, 158, 162, 168, 191
Bronchial associated lymphoid tissue (BALT),
    48
Bronchoconstriction, 147–148, 150

Brookes and Lawley, 155
Burgdorfer, 187
Burkitt's lymphoma, 158
Business model, 233

C
C3 convertase, 27
C5 convertase, 28
Caffeine, 214
Calcitonin, 62, 75
Calcitonin gene related peptide (CGRP), 29, 85,
    88, 90–91
Calcium, 74
California encephalitis virus, 186
Caloric intake, 217
Cancer, 3, 111, 114, 135, 154, 167–168,
    175–176, 191, 212, 217–218, 220, 227
    breast, 158, 162, 168, 191
    Burkitt's lymphoma, 158
    cervix, 163
    defenses against, 158–159
    diet, 158
    hormones, 158
    infection, 158
    melanoma, 161–162
    oncogenes, 156
    pancreas, 160
    psychosocial stress, 159–164
    toxins, 157
    tumor supressor genes, 156, 175
Candida albicans, 137, 163
Cannon, Walter, 17
Capsid, 136
Carcinogens, 157–158
Carcinogenic chemicals, 155
Cardiorespiratory arrest, 143
Cardiovascular disease, 99, 113, 138–139, 150,
    220
Carrier state, 135
Catatonia, 200
Catecholamines, 62, 74, 129, 217, 220
CD3+, 37–38
CD4+, 37–38, 77, 126–127, 136, 142, 163, 166,
    169, 174, 176, 202, 216
CD4 + CD29+, 202
CD4 + CD45+, 202
CD8+, 37–38, 66, 77, 126–128, 146, 163, 166,
    202, 216
CD28, 166
CD40L, 166

CD40R, 166
CD56 marker, 162
CD80, 166
CD86, 166
Cell death, 155, 158, 175
Cell growth, 155
Cell-adhesion molecules, 42, 46, 97, 138, 174, 182
Cell-mediated immunity, 24, 40, 54, 67, 71, 74, 76, 97, 126–128, 148, 150, 166, 169, 172, 175–176, 183, 213, 218–219, 227
Central nervous system (CNS), 12, 92, 123, 146–147, 165, 167–168, 180–185, 187–190, 192–193, 227
Central nervous system and immune modulation
  amygdala, 60, 78, 94–95, 99
  anterior cingulate cortex, 94
  anterior hypathalamus, 65, 95
  area postrema, 60, 88, 94
  basal ganglia, 78, 199
  bed nucleus of the stria terminalis, 94–95
  cerebellum, 78, 198
  dorsal motor nucleus of the vagus, 92
  hippocampus, 60, 70, 78, 94–95, 123, 201, 227
  hypothalamus, 17, 59, 64, 73, 78, 93, 192
  insular cortex, 94, 99
  intermediolateral column of the spinal cord, 92
  midline thalamic nuclei, 92
  neocortex, 78, 94, 96, 181
  nucleus ambiguus, 92
  nucleus basalis, 95
  nucleus of the tractus solitarius, 92, 96
  organum vasculosom of the lamina terminalis, 60, 73, 88, 94, 192
  parabrachial nuclei, 92, 95
  paravertricular nucleus of the hypothalamus, 60, 72–73, 92, 96
  periaqueductal grey, 84, 92
  reticular formation, 84, 94–95
  septal nuclei, 60, 95
  subfornical organ, 60, 73, 88, 94
  substantia nigra, 95
  thalamus, 60, 78, 92–93
  ventral medulla, 92
  ventromedial nucleus of the hypothalamus, 96
  ventromedial prefontal cortex, 94
Central tolerance, 164
Cereals, 143
Cerebellum, 198
Cerebrospinal fluid (CSF), 97, 182, 186, 201–202

Cervical cancer, 163
Cervical dysplasia, 163
Chachectin (TNF-α), 44
Change in residence, 107
Chaos theorists, 229
Chaotic, 140
Chemical exposure, 103
Chemoattractant cytokines, 46
Chemokines, 42, 46
Chemotaxis, 41, 50, 91, 216
Chemotherapy, 10, 155
Chinese, 12
Chlamydia pneumonial, 138
Chlamydias, 136
Chorea, 198
Choroid plexus, 190
Christian, 11
Chromosone 6, 31–32
Chromosones, 158
Chronic diseases, 139
Chronic fatigue syndrome, 4, 188, 194–195, 214
Chronic hostility, 139
Circadian rhythms, 62
Circumventricular organs, 60, 73, 88, 94, 192
Cirrhosis, 165
Classical conditioning, 18, 98–99, 212, 221
Classical pathway, 28
Clergy, 208
Climate, 230
Clinical decision-making, 233
Clonal expansion, 42
Clonal selection theory, 164–165
Cluster of differentiation (CD), 37
CNS surveillance, 182
Coca and Cooke, 143
Cocci, 136–137
Cognition, 82, 93, 173, 180
Cohen, Irun, 165
Cohen, Sheldon, 142
Cohort, 130
Cold exposure, 195
Collateral damage, 135
Colonoscopy, 122
Colony stimulating factors (CSFs), 44, 182–183
  eosinophil, 44
  granulocyte-monocyte, 44
  macrophage, 44
  megakaryocytes, 44
Coma, 188, 203
Competition, 215

Complement, 26, 29, 50, 53, 54, 166–167, 169, 176, 182–183, 216, 218
Complexity, 130, 228
Concanavalin-A (Con A), 52
Conditioned stimulus (CS), 98–99
Conditioning, 98–99, 208, 211
Conscientiousness, 100
Conscious awareness, 94
Consciousness, 112
Conservation of complexity, 228
Constant segment, 36
Contextual change, 3, 103–115, 129, 141, 150, 188, 227
Cooperative, 163
Coping, 105, 141, 162
Copper, 218–219
Corneal transplants, 189
Corticosteroid, 188, 190, 216
Corticosteroid binding globulin, 69, 120
Corticotrophs, 68
Corticotropin, 59
Corticotropin releasing hormone (CRH), 59, 67–68, 85, 147, 169
Cortisol, 74–75, 120–121, 130, 142–143, 160, 163, 171, 174–175, 216–217
Cost-containment, 232
Cost-effective, 232
Counter-shock reaction, 104
Cousins, Norman, 168
Coxsakie virus B, 167
C-reactive protein, 138
Creutzfeldt-Jakob disease, 136, 186
Crohn's disease, 166
Crowding, 105
Culture, 105, 108, 157, 208, 211, 231
Curse, 211
Cyclosporine, 67
Cyclic-adenosine monophosphate (cAMP), 123
Cytokines, 29, 42, 49, 54, 63, 70, 77, 89, 94, 96, 111, 127, 138, 142, 166, 168–169, 174, 176, 180, 183, 191–192, 195, 197, 201–202, 216–217, 220
  antiinflamatory, 43–44
  chemoattractant, 46
  leukocyte inhibitory factor, 45
  macrophage inflammatory proteins, 45
  migratory inhibitory factor, 46
  proinflammatory, 45
  see also colony stimulating factor
  see also interferons
  see also interleukins
  see also transforming growth factor
  see also tumor neurosis factor
Cytomegalovirus (CMV), 126, 138, 187, 201
Cytoplasm, 136
Cytotoxic T-cells, 39, 158, 166, 172
Cytotoxicity, 34, 41, 66

D

D2-receptors, 66
D8/17 marker, 199
Daily hassles, 124, 127
Davidson, Richard, 100
Death, 8, 135, 171, 187, 211
Death of spouse, 106–109
Decision making, 119
Delayed hypersensitivity skin test (DHST), 52
Delayed type hypersensitivity reaction, 41, 72, 90, 95, 97, 126–127, 202, 209–210, 216
Delirium, 191
Delusions, 191, 199
Dementia, 97, 186–187
Dementia praecox, 199
Dendritic cells, 26, 30, 43–44, 52, 90, 145
Deoxyribonucleic acid (DNA), 111, 136, 155–156, 158, 175, 188, 219, 228
Depression, 4, 124, 139, 141, 160, 171, 173–176, 187, 191, 194–196, 201, 214, 233
Detached, 140
Diagnostic Manual of the American Psychiatric Association (DSM-IV), 200
Diarrhea, 143
Diet, 4, 142, 158, 211, 219–220, 222, 230
DiGeorge syndrome, 76
Dimensions of context, 106
Disclosure of traumatic events, 214
Disease, 1, 4, 113, 184, 208, 211, 213, 217, 222, 225, 227–233
  early ideas, 10–14
  empirical approaches, 14–15
  magic, 10–11
  philosophy, 11–14
  prevention, 230
  religion, 10–11
  science, 15–19
  social organization, 9–10
Diseases of adaptation, 105
Disulfide bonds, 35
Diversity segment, 36
Divorce, 106–109, 126, 171–172

Dizygotic twins, 144
DNA repair, 155, 158, 162, 175
Doctor, 208, 212, 232
Dopamine, 66, 70, 72–73, 95
Dopamine receptors, 87
Dorsal motor nucleus of the vagus, 92
Dorsal thalamus, 84
Dorsal root ganglia (DRG), 84
Dosas, 13
Double-blind random assignment methodology, 210
Drug abusers, 141
Drugs, 144, 162, 168, 176, 190, 213
Dunbar, Helen Flanders, 18
Dust mites, 143
Dynamics of disease, 230

E

Earthquake, 118, 124, 127–128, 130
Economics, 233
Eczema, 143–145
Edema, 145, 147
Education, 172
Efferent lymphatics, 47
Eggs, 143
Egyptian, 10
Egyptian mummies, 154
Ehrlich, Paul, 16, 164
Eicosanoids, 220
Electric shock, 2, 119, 124
Electroencephalographic (EEG) abnormalities, 173
Electrolytes, 104
Emotion, 82, 93, 228
Encephalitis epidemic, 197
Encephalopathies, 186
Endemic, 134
Endocrine, 62, 119–122, 226
Endocrine system, 2, 58, 103
  adrenal glands, 62
  anterior pituitary, 60–61
  hypothalamus, 59–60
  islets of Langerhans, 62
  ovaries, 62
  parathyroid glands, 62
  pineal gland, 62
  posterior pituitary, 61–62
  testes, 62
  thymus gland, 63
  thyroid gland, 62

Endogeneous opioids, 66
β-Endorphin, 61, 63, 68–69, 75–76, 88, 120, 216
Endothilial cells, 43–44, 58, 91, 138, 174, 181–182, 187, 192
Endothelial leukocyte adhesion molecule-1 (ELAM-1), 46
Engel, George, 18
Enkephalins, 85, 88
Enmeshed, 140
Enteric ganglia, 86
Enteric nervous system (ENS), 82, 85–86
Entering school, 127
Enzymes, 51, 111, 114, 136, 146, 222
Enzyme-linked immunosorbent assay (ELISA), 51
Eosinophils, 25, 28–29, 44, 91, 124, 145
Epidemics, 134–135
Epidemiological studies, 110, 200, 213
Epidemiology, 233
Epinephrine, 62, 74, 122, 142, 147, 216
Epitopes, 31
Epstein-Barr Virus (EBV), 126, 158, 167, 187, 195
Erythrocytes, 167
Essential fatty acids, 220
Essential nutrients, 218
Estradiol, 62, 70–71, 121, 169
Estrogens, 66, 68, 169, 174
Eukaryotes, 136–137
Evil-eye, 11
Evolutionary principles, 112, 147
Evolve, 229
Ewald, Paul, 139
Exercise, 4, 103, 124, 215–217, 222
Excitatory amino acids, 70
Exhaustion phase, 104
Exocytosis, 136
Expanding clone, 155
Expectations, 4, 208–209, 211–212, 221–222, 228, 232
Exposure, 230
Expression of emotion, 208, 213
Extravasation, 42, 46
Extraversion, 100–101

F

Fábrega, Horacio, 8
Fab region, 35
Faleide, 149

Family atmosphere, 140
Family history, 171, 176, 202
Fantasy, 109
Fawzy, F., 162
Fatalistic, 141
Fatigue, 111, 139, 162, 187, 191, 194–195
Fc region, 35, 172
Fessel, 19
Fever, 88, 138, 189, 192, 194
Fibrinogen, 29, 138
Fibroblasts, 43–45, 187
Fibromyalgia (FM), 194–195, 214
Fighting spirit, 161, 175
Financial gain, 210, 231, 233
Financial liability, 232
Financially responsible, 233
Flare and wheal response, 209
Flat affect, 200
Flatworms, 137
Flora, 197
Flow cytometry, 52
Focal neurologic defects, 185
Follicle stimulating hormone (FSH), 60, 70–71, 121
Food allergy, 143–144, 146, 191, 197–198
Foreclosure, 107
Forgetfulness, 186
Fox, Bernard, 160–161
Francostoro, 134
Free fatty acids, 65
Free-radical scavenger, 75
Frontal cortex, 201
Fungi, 137, 184
Funkenstein, 149

G

Gabaergic neurons, 123
Galanin (GAL), 65, 85, 88
Galen, 13, 17, 155, 159
Gamma amino butyric acid (GABA), 65, 68, 70
Ganglia, 85–86
Gaskill, Walter H., 17
Gastric ulcers, 138
Gastrin releasing peptide, 88
Gastritis, 137
Gastroenteritis, 137
Gastrointestinal disorders, 139
Gastrointestinal tract, 68, 82, 85–86, 146, 186, 192, 196–197, 202

Gender, 169, 174, 230
Gene, 111, 145, 155–156, 158, 164, 169, 171, 175, 228
General adaptation syndrome, 104, 194
General paresis, 184
Genetic endowment, 104
Genetic factors, 99, 169, 230
Genetic mutations, 155
Genetic profiles, 230
Genetic susceptibility, 228
Genome, 111, 114, 230
Genomic expression, 111, 227
Germ-theory of disease, 10, 15
Gerstmann-Strausser-Scheinker, 136
Gestation, 78, 196
Glial cells, 58, 192
y-Globulin group, 35
Glomerulonephritis, 165
Glossopharyngeal nerve (IX), 73
Glucagon, 62, 65, 74–75
Glucocorticoid, 61, 66, 68–69, 96, 123, 148, 150, 169, 174, 216, 220
Glucose, 74, 104, 163, 220
Glutamate, 70
Glutamine, 218
Glycine, 185
Glycoprotein, 60
Golub, Edward, 9, 17
Gonadotropins, 70–71
Gonads, 62, 73, 76
Gonorrhea, 138
Goodwin, P., 161–162
Gp 120, 190
Gram stain, 136
Gram-negative, 136, 168
Gram-positive, 136
Granulocyte-macrophage colony stimulating factors (GM-CSFs), 66, 69
Granulocytes, 25, 28, 65
Granuloma, 138
Granulopoiesis, 191
Grave's disease, 165–166, 170
Greaves, Mel, 138, 156
Greek, 154
Grooming, 121, 194
Growth hormone (GH), 60, 63–66, 74–77, 121, 129–130, 216, 220
Growth hormone releasing hormone (GHRH), 60, 64
Guillain-Barré syndrome, 186, 188–189

Guru, 208
Gut associated lymphoid tissue (GALT), 48, 50,
    128, 192

H

Haemophilus influenza, 184
Hall, Stephen, 159
Hallucinations, 188, 191, 199
Hand washing, 198
Hapten, 144
Hardiness, 141
Harvey, William, 15
Hashimoto's thyroiditis, 164–165
Hay fever, 143–145
Headaches, 145, 189
Health, 1, 114, 184, 207–208, 213–215, 218, 221,
    225–226, 228, 230–233
    early ideas, 10–14
    empirical approaches, 14–15
    magic, 10–11
    philosophy, 11–14
    religion, 10–11
    science, 15–19
    social organization, 9–10
Health care, 4, 225, 231–233
Health education, 213, 232
Health habits, 142, 176
Health maintenance, 124, 228
Health plans, 232–233
Health state space, 229–231
Heart rate, 212
Heath, Robert, 19, 202
Heavy chains, 35
Heavy-chain class switching, 40
Helicobacter pylori, 137–139
Helminths, 137
Helper T-cells, 136, 145, 166, 172, 174
Helpless, 163
Hematogenous route, 181
Hematopoiesis, 64
Hematopoietic stem cells (HSC), 24–25
Hemolytic anemia, 164–165
A β-Hemolytic streptococcus, 167, 185, 198
Heparin, 28
Hepatitis-B virus, 158
Hepatocytes, 158
Herd immunity, 135
Herpes Simplex virus (HSV), 126, 140, 181,
    187–188, 195–196, 198, 201, 213–214
Herpes virus – 6, 187

Herpes virus – 7, 187
Herpes virus – 8, 187
High endothelial venule cells (HEV), 47
High-external-locus-of-control 125
Hindu, 10, 155
Hippocampus, 60, 70, 94–95, 123, 201, 227
Hippocrates, 13, 155, 159
Hirata-Hibi, 19
Histamine, 28–29, 69, 75, 145, 209
HIV infection, 38
Hives, 143
HLA B27, 168
HLA class I molecules, 31–33, 53
HLA class II molecules, 31–33, 169, 182, 202
HLA DQ2, 169
HLA DQ8, 169
HLA DR2, 174
Holism, 9, 232
Holmes and Rahe, 106–108
Homeopathy, 10
Homo erectus, 154
Hopelessness, 141, 161, 163
Hormones, 2, 17, 63, 111, 117, 129–130, 158,
    174, 215, 220, 226
    adrenaline, 62, 122
    adrenocorticotropin hormone, 59–60, 67–70
    arginine vasopressin, 60, 68, 73
    calcitonin, 62
    corticotropin releasing hormone, 59, 67–70
    estradiol, 62, 70–71
    follicular stimulating hormone, 60, 70–71
    glucagon, 62
    glucocorticoids, 62, 69–70
    gonadotiopin releasing hormone (GnRH),
        60–61, 70–71, 121
    growth hormone, 60–61, 64–66
    growth hormone releasing hormone, 60–61
    human chorionic gonadotropin hormone, 71
    insulin, 62, 74
    insulin-like growth factor-1, 66
    luteinizing hormone, 60–61, 70–71, 76
    α-melanocyte stimulating hormone, 61, 68–69
    melatonin, 62, 75
    mineralocorticoids, 62
    oxytocin, 60–61, 73–74
    parathyroid hormone, 62, 74–75
    progesterone, 62, 71–72
    prolactin, 60–61, 66–67
    somatotropin release inhibiting hormone,
        64–65

testosterone, 62, 71–72
thyroid stimulating hormone, 60–61, 72
thyrotropin releasing hormone, 60–61, 72
thyroxine, 62, 72
triiodothyronine, 62, 72
Horror autotoxicus, 164
Hostility, 101, 171
Hughes, 18
Human chorionic gonadotropin hormone (hCG), 71
Human genome, 230
Human immuno deficiency virus (HIV), 112, 140–142, 150, 163, 189, 190, 201
Human Leukocyte Antigen (HLA) system, 31, 176, 182, 231
  Locus A, 31
  Locus B, 31
  Locus C, 31
  Locus DP, 31
  Locus DQ, 31
  Locus DR, 31
Human papilloma virus (HPV), 163
Humoral immunity, 24, 40, 54, 125–126, 128, 146, 148, 166, 169, 172, 176, 218
Humors, 155
Hurricane, 124, 130
Hydrophobia, 189
Hyperactivity, 185, 199
Hyperparathyrodism, 75
Hyperphagia, 185
Hypersomnia, 195
Hyperthyroidism, 165–166, 170
Hypnosis, 4, 208–210, 213, 215, 221
Hypoglycemia, 65
Hypophysectomy, 64, 95
Hypophysiotropic, 59
Hypophysis, 60
Hypotension, 143
Hypothalamic–pituitary–adrenal axis (HPA axis), 46, 68, 129–130
Hypothalamic–pituitary–gonadal axis (HPG axis), 70
Hypothalamus, 17, 59, 64, 73, 93, 192
Hypothyroidism, 165

I

Idiotype, 49
Idiotype/anti-idiotype interactions, 49
Ignorance, 232
IL-1RA, 45, 63, 217

Imagination, 208
Immediate hypersensitivity reactions, 209
Immune activity, 180, 183–184, 187, 191–192, 194, 197, 203, 208, 211, 215, 221, 225–226, 231
Immune complexes, 166–167, 176, 190
Immune function, 141, 214–215, 217–218, 221, 226
Immune function enhancement
  antioxidants, 219–220
  beliefs, 208–209
  emotional expression, 213–214
  exercise, 215–217
  hypnosis, 209–210
  nutrition, 217–219
  placebo, 210
  relaxation, 214–215
  sleep, 214–215
  social activity, 213
  suggestion, 208–209
Immune function measures, 51–53
Immune system, 2, 82, 103, 111, 117, 129, 140, 149, 156, 158, 162–164, 175–176, 183, 186, 191, 193, 195, 207–209, 214–215, 217–219, 221, 225–227, 230–231
  antibody diversity, 35–37
  antibody structure, 35–37
  antigen presentation, 32–34, 70
  basophils, 128
  bone marrow, 46–48
  cell-adhesion molecules, 46
  cell-mediated responses, 37–42
  chemokines, 42
  colony stimulating factors, 44–45
  complement, 25–28
  cytokines, 42–46
  dendritic cells, 30
  humoral responses, 35–37
  inflammation, 28–31
  interleukins, 43–44
  leukocytes, 25–26
  leukocyte traffic, 42, 47, 69
  lymphatic system, 46–48
  lymph nodes, 46–48
  lymphoid organs, 46–48
  B-lymphocytes, 35–37
  T-lymphocytes, 37–38, 42
  macrophages, 30
  CD-markers, 37–38
  monocytes, 30

Immune system (*continued*)
  mucosa associated lymphoid tissue, 46–48
  natural killer cells, 34
  neutrophils, 28–29
  T-receptor diversity, 37–38
  T-receptor structure, 37–38
  self/non-self discrimination, 24
  spleen, 46–48
  thymus, 46–48
Immunization, 95, 214
Immunodeficiency, 64, 218
Immunofluorescence assay, 52
Immunogenicity, 135
Immunoglobulins (Ig), 35, 216, 218
  disulfide bonds, 35
  Fab region, 35
  Fc region, 35, 172
  heavy chains, 35–36
  IgA, 36, 48, 50–54, 91, 126, 128
  IgD, 36, 50
  IgE, 28, 36, 52, 54, 91, 143, 145–146, 148, 150
  IgG, 36, 50–52, 54, 183, 202
  IgM, 36, 50, 52
  light chains, 35–36
Immunoglobulin-supergene family, 46
Immunotherapies, 159
Inanimate objects, 229
Inattentiveness, 186, 199
Incantations, 11
Incubation period, 135
Individual characteristics, 106, 118
Infection, 64, 72, 97, 103, 105, 135, 145, 150,
    158, 167, 175–176, 180–181, 184–185, 189,
    191, 193–194, 196–198, 200, 202–203, 209,
    214, 217–219, 221–222, 233
Infection, immune activity and behavior
  aggression, 185, 189
  anorexia, 187, 189, 191, 195
  anxiety, 188, 195, 198
  apathy, 200
  aphasia, 186
  ataxia, 186
  attention-deficit-hyperactivity disorder,
    197–199
  autism, 196–198, 201
  bipolar disorder, 191, 201, 203
  catatonia, 200
  chronic fatigue syndrome, 188, 194–195
  coma, 200
  delirium, 191

  dementia, 186–187
  depression, 187, 191, 194–196, 201
  fibromyalgia, 194–195
  flat affect, 200
  forgetfulness, 186
  Guillain-Barré syndrome, 186, 188–189
  hallucinations, 188, 191, 199
  hydrophobia, 189
  hypersomnia, 195
  inattention, 186, 199
  insomnia, 195
  irritability, 186–187, 189, 191, 199
  lethargy, 185, 194
  Lyme disease, 184, 187
  malaise, 191, 194
  mental retardation, 185, 196–197, 201
  neurasthenia, 194
  nightmares, 198
  obsessive compulsive disorder, 186–187,
    198–199
  paranoia, 191
  pervasive developmental disorders, 196
  photophobia, 187
  post encephalitic behavior, 197
  schizophrenia, 187–188, 199–203
  sickness behavior, 193–196
  sleeplessness, 186
  social withdrawal, 200
  Sydenham's chorea, 185, 198
  Tourette's syndrome, 198–199
Infectious disease, 3, 113, 134, 139, 141,
    149–150, 154, 200, 203, 214, 230
  acquired immune deficiency syndrome,
    137–138, 140–141, 142
  Creutzfeldt-Jakob disease, 136
  encephalitis, 184
  gastritis, 137
  gastroenteritis, 137
  Gerstmann-Strausser-Scheinker, 136
  gonorrhea, 138
  influenza, 140
  kuru, 136
  Lyme disease, 137, 184, 187
  malaria, 137, 194
  meningitis, 184
  mononucleosis, 140, 202
  peptic ulcers, 137, 139, 150
  pharyngitis, 137
  plague, 134
  pneumocystis carinii, 137

poliomyelitis, 138
sleeping sickness, 137
smallpox, 134
syphilis, 194
toxoplasmosis, 137
tuberculosis, 138
upper respiratory colds, 140, 142
Infectious disease role in
allergy, 145–147
autoimmune disorders, 167–168
cancer, 158
cardiovascular disease, 138–139
gastrointestinal disorders, 139
neurologic disorders, 185–190
psychiatric disorders, 194–203
Infectious mononucleosis, 140, 202
Infectivity, 135
Inflammation, 16, 19, 24, 28–29, 50, 69–70, 88,
    91, 126, 138, 143–145, 150, 166, 172, 176,
    183, 185, 187–188, 190, 193, 197–198, 203,
    209, 212, 214–215, 217, 219–222
Inflammatory bowel disorders, 166, 173
Inflammatory process, 29
Influenza, 110, 140, 185, 200
Inhibited-need-for-power, 125
Innate heat, 14
Inoculation, 96
Insanity, 200
Insects, 109
Insomnia, 195
Insular cortex, 94, 99
Insulin, 62, 74
Insulin dependent diabetes, 165–167, 169, 173
Insulin-like growth factor-1 (IGF-1), 63, 65–66,
    77
Intergrins, 46
Interdigitating cells, 30
Interferons (INF), 217
    INF-α, 44–45, 127, 162, 191, 195
    INF-β, 44–45
    INF-γ, 29, 39, 44–45, 67–69, 71, 74, 90, 166,
        182–183
Interleukins (ILs), 45
    IL-1, 29, 43, 45, 63, 67–70, 88–89, 91, 96,
        182, 192–193, 217, 220
    IL-2, 39, 43, 45, 51, 63, 67, 69, 71–72, 74, 91,
        127, 191, 195, 197, 202
    IL-3, 43, 45, 66, 69
    IL-4, 29, 39, 43, 45, 69, 71, 75, 127, 166, 169
    IL-5, 39, 43, 45

IL-6, 29, 43, 45, 63, 68–70, 91, 142–143, 164,
    182–183, 193, 195, 217
IL-7, 43, 45, 91
IL-8, 29, 43, 166
IL-9, 43, 45
IL-10, 39, 43, 49, 71, 90, 127, 166, 169, 217
IL-11, 43
IL-12, 30, 39, 43, 69, 90, 166
IL-13, 39, 43
IL-14, 44
IL-15, 44–45
IL-16, 44
IL-17, 44
IL-18, 44
Intermediolateral column, 92
Internal milieu, 113
International Leukocyte Workshops, 37
Interstitial dendritic cells, 30
Intracellular adhesion molecule-1 (ICAM-1), 46
Intracellular messengers, 25
Intracranial tumors, 97
Intravenous IgG, 188
Introversion, 124–125, 142, 171
In utero, 217
Investors, 233
Iodine, 168
Ionizing radiation, 155
Iron, 218
Irritability, 186–187, 189, 191, 199
Islet of Langerhans, 62, 74
Isolation, 105
Isotypes, 36, 49
Itching, 143

J
Jail term, 106–107
Jenner, Edward, 16
Jing, 12
Job stress, 12
Joining segment, 36
Justice, 163

K
Kanner, 196
Kapha, 13
Karcinos, 159
Kinins, 29
Kitusata, Shibaraburo, 16
Klopfer, Bruno, 212
Kraepelin, Emil, 199–200

Krebiozen, 212
Kuru, 136, 186

L

Lack of sleep, 195
Langerhans cells, 30
Langley, John Newport, 17
Language disabilities, 97, 187, 196
Large granular lymphocytes, 162
Lassitude, 194
Latent infection, 113, 135, 140, 195
Latent viruses, 126, 188, 201
Lateral ventricles, 201
Layers of context, 111
Lead poisoning, 197
Learning, 98–99
Learning deficits, 64
LeDron, Henri Francois, 155
Left-handedness, 97
Left-hemisphere, 96, 214, 221
Left-subclavian vein, 47
Legallois, Jean-Cesar, 17
Lethargy, 185, 194
Leukemia, 157
Leukocytes, 25, 31, 63, 68, 70, 82, 97, 101, 128,
    138, 145, 157, 159, 166, 181, 210, 214–215,
    218, 227, 231
Leukocyte traffic, 69, 89
Leukocytosis, 90, 215–216
Leukotrienes, 28–29, 69, 91, 145, 220
Levy, Sandra, 162
Liability, 232–233
Life events, 106, 118, 140, 142, 150, 159, 161,
    163, 170, 175, 188
Light chains, 35
Light/dark cycle, 62, 75
Limbic structures, 60, 181
Limbic system, 94–95, 99
Lipid mediators, 29
Lipids, 104, 138, 218, 220
Lipocortin-1, 69
Lipoprotein, 136
Lister, Joseph, 15
Listeria monocytogenes, 137
Local ecology, 108
London Cancer Hospital, 159
London Epidemic of 1665, 134
Loneliness, 124, 211
Loss, 130, 171, 173
Loss of synapses, 201

Louse, 137
Luteinizing hormone (LH), 60, 70–71, 121
Lyme Disease, 110, 137, 184, 187
Lymphatic system, 182
Lymphocyte proliferation, 31, 64, 66, 68, 91, 96,
    101, 142, 195, 213–215
Lymphocytes, 25, 30–31, 43–45, 68, 72, 90, 95,
    128, 164, 166, 168, 180, 183, 186, 202, 210,
    213–214, 216, 218
Lymphocyte function associated antigen-1 (LAF-
    1), 46
Lymph nodes, 33, 47–49, 72, 128
Lymphocytosis, 216
Lymphoid cancer, 159, 212
Lymphoid follicle, 35
Lymphoma, 164
Lymphotoxin (TNF-β), 44

M

Macrophages, 26, 30, 43–45, 50, 52, 65, 67, 138,
    145–146, 167, 183, 187
Magnesium, 74, 219
Maier and Watkins, 192–193
Major histocompatability complex (MHC), 31
Malaise, 111, 138, 191, 194
Malaria, 110, 137, 194
Malignancy, 3, 53, 66, 113, 154, 158, 162–163,
    180, 190–191, 200, 220, 230
Malignant disease, 135
Malnutrition, 218
Mammotrophs, 66
Managed-care companies, 233
Mania, 187
Marine organisms, 143
Marital discord, 127
Marriage, 106–109
Massage, 213
Mast cell degranulation, 28, 54, 89, 91, 147, 150
Mast cells, 25, 28, 43, 89, 145–146, 150, 220
Mating behavior, 78
MB-35 peptide, 76
Mecholyl, 149
Medicines, 228
Media coverage, 212
Median eminence, 59, 73
Medical assistance, 113, 163, 176
Megakaryocytes, 25
Melancholia, 155
α-Melanocyte stimulating hormone (α-MSH),
    61, 68–69

Melanoma, 161–162
Melatonin, 62, 75
Membrane attack complex, 28
Memory cells, 40, 52, 216
Menninger, Karl, 200
Menopause, 169, 173
Mental activity, 108, 113, 127, 227, 231
Mental arithmetic, 113, 119, 124
Mental attitude, 228
Mental disorders, 208
Mental retardation, 185, 196–197, 201
Mercury, 168
Mesenteric/celiac ganglia, 86
Meta-analysis, 161
Metastasis, 159, 168, 191
Metchnikoff, Elie, 16
Methylation, 156
Microenvironments, 4, 30, 226–228
Microganglia, 87
Microglia, 97, 174, 182, 189
Microorganisms, 134–135, 149, 219
Midbrain, 62, 70
Migraine, 146
Migration inhibitory factor (MIF), 44, 46
Milieu intérieur, 16
Milk, 143, 146
Mineralocorticoids, 62
Minerals, 218
Mini-space, 230
Mini-tumors, 138
Minnesota Multiphasic Personality Inventory
    (MMPI), 160
Missile attack, 127
Mitogens, 52, 68, 74, 95, 126, 195, 202,
    210, 216
Mobility, 108–110
Molds, 137
Molecular mimicry, 114, 138, 167–168
Monoaminergic neurons, 123
Monocytes, 25, 30, 43–45, 68, 124, 189, 216
Mononuclear leukocytes, 25, 67, 189
Monozygotic twins, 144, 169–172, 176
Morning after pill, 175
Motivation, 82, 93
Mucosa associated lymphoid tissue (MALT), 33,
    47–48, 50, 87
Mucus hypersecretion, 147
Multiorgan autoimmunity, 167
Multiple sclerosis, 110, 165–166, 173–174, 203
Mumps measles rubella (MMR) vaccine, 197

Muscarinic receptors, 87
Mutations, 111, 155–156
Myalgias, 195
Myasthenia gravis, 165–166
MYC gene, 156
Mycobacterium tuberculosis, 52, 138
Mycoplasmas, 136
Myelin, 182
Myenteric plexus, 86
Myocardium, 167

N

Natural healing process, 232
Natural killer (NK) cells, 25, 31, 34, 44, 51–54,
    64, 66, 71, 75, 89, 95–96, 101, 124–125,
    128, 142, 162, 195, 213–216
Naturopathy, 10
Negative affectivity, 96, 124, 149, 174, 215
Negative emotion, 175, 213–214, 221
Negative placebo, 211
Negative symptoms, 200, 202
Neisseria meningitis, 184
Neisseria gonorrheal, 137
Neocortex, 94, 96, 181
Neocortical-amygdalar circuits, 100
Neoplastic disease, 154
Nephelometry, 52
Nerve growth factor (NGF), 97
NEU gene, 156
Neural development, 201
Neural networks, 100
Neurasthenia, 194
Neuritic plaques, 186
Neurodegenerative disorders, 180, 186
Neuroendocrine pathways, 130, 225
Neurofibrillary tangles, 186
Neurogenic inflammation, 91, 147, 209
Neuroglia, 67
Neuroimaging studies, 198, 201
Neuroinvasive, 181
Neurokinin A, 147
Neuroleptics, 67, 160, 191, 202
Neurological disorders, 181, 185–186
Neurons, 58, 183, 185, 188, 190, 192
Neuropeptide Y (NPY), 68, 85, 88, 90–91
Neuropeptides, 60, 65, 85, 111
    calcitonin gene related peptide, 29, 85, 88,
        90–91
    corticotropin releasing hormone, 59, 67–68,
        85, 147, 169

Neuropeptides (*continued*)
  β-endorphin, 61, 62, 68–69, 75–76, 88, 120, 216
  enkephalins, 85, 88
  galanin, 65, 85, 88
  gastrin releasing peptide, 88
  neurokinin A, 147
  neuropeptide Y, 68, 85, 88, 90–91
  peptide histidine isoleucine amide, 88
  pituitary adenylate cyclase activating polypeptide, 88
  somatostatin, 29, 60, 65, 74, 85, 88, 90
  substance P, 29, 68, 85, 88, 90, 122, 147
  vasoactive intestinal peptide, 20, 65, 66, 85, 88, 90–91, 147
Neurophysin, 73
Neurosyphilis, 184, 187
Neuroticism, 100
Neurotoxicity, 189
Neurotoxins, 190, 196
Neurotransmitters, 60, 65, 85, 96, 111, 117, 130, 174, 185, 191, 193
  acetylcholine, 68, 74, 85, 95, 122, 147, 166, 185, 189
  dopamine, 66, 70, 72–73, 95
  gamma-amino-butyric acid, 65, 68, 70
  glutamate, 70
  N-methyl-D-aspartate, 70
  nitric oxide, 29, 68, 70, 89, 91, 147
  noradrenaline, see norepinephrine
  serotonin, 28–29, 68, 95, 123
Neurotrophic, 92, 181, 187
Neurotrophic factors, 182, 190
Neurovirulent, 181
Neutral proteases and hydrolases, 28
Neutrophils, 25, 28–29, 50, 91, 210, 216, 218–219
Nightmares, 198
Nitric oxide (NO), 29, 68, 70, 89, 91, 147
N-methyl-D-aspartate (NMDA), 70
Nocebo effect, 209, 211, 232
Nociceptive neurons, 147
Nociceptors, 83
Non-atopic individuals, 145
Non-living environment, 110–111
Norepinephrine, 62, 65, 68, 72–73, 75, 94, 96, 122, 142, 216
Nucleic acids, 218
Nucleus ambiguus, 92
Nucleus basilis, 95

Nuclei of the solitary tracti (NTS), 92, 96
Nutrients, 220
Nutrition, 217, 221–222, 227–228
Nutritional choices, 208
Nuts, 143

O

Obesity, 217
Obligate intracellular parasites, 136
Obsessive-compulsive disorder (OCD), 4, 19, 186–187, 198–199
6-OHDA, 90
Olfactory mucosa, 188
Oligodendrocytes, 182
Omega-3 (w-3) fatty acid, 220–221
Omega-6 (w-6) fatty acid, 220
Oncogenes, 156
Oncostatic effect, 75
Openness to experience, 100
Opioid peptides, 69, 74
Optimistic, 141, 161, 171
Orbito-frontal cortex, 188
Organ, 111, 230
Organ specific autoimmune disorder, 164
Organ transplant, 163
Organum vasculosum of the lamina terminalis (OVLT), 60, 73, 88, 94, 192
Osler, 190
Osmoreceptors, 73
Osteopathy, 10
Other life forms, 106, 109
Ovarian cancer, 168, 191
Ovaries, 62, 70
Over-arousal, 185
Oxidative metabolism, 222
Oxygen free radicals, 158, 175, 219–220
Oxytocin, 59, 73, 121
Ozone, 220

P

P53 gene, 156
Pain, 124, 195, 209, 212, 214, 233
Pain management, 213
Pancreatic cancer, 160
Panic, 187
Parabrachial nuclei, 92, 95
Parachute jumping, 118, 121–122
Paracrine, 42, 63, 74
Paraganglia, 193
Parakritic, 13

Paralysis, 188–189
Paraneoplastic syndromes, 168, 191
Paranoia, 191
Parasites, 28, 146, 184
Parasitic infections, 145–146, 150
Parasympathetic nervous system (PSNS), 17, 49, 82, 85–86, 122, 147
Parathyrodectomy, 75
Parathyroid hormone (PTH), 62, 74–75
Paraventricular nucleus (PVN), 60, 72–73, 92, 96
Paravertebral ganglia, 85–86, 99
Paretic side, 97
Parkinson's disease, 186, 203
Passive, 163
Pasteur, Louis, 15–16
Pathogenicity, 135
Pavlov, Ivan, 17, 98
PCR technique, 201
Peanuts, 143
Pediatric autoimmune neuropsychiatric disorders (PANDAS), 199
Pennebaker, James, 214
Peptic ulcers, 137, 139, 150
Peptide histidine isoleucine amide (PHI), 88
Peptides, 68, 117, 122, 129–130, 147, 174, 196
Perfectionistic, 171
Periaqueductal grey matter, 84, 92
Periarteriolar lymphocyte sheath (PALS), 48
Peripheral nervous system (PNS), 2, 78, 82, 122, 226
Peripheral nervous system and immune modulation
    autonomic nervous system, 82, 84–85, 88, 90
        parasympathetic branch, 82, 85–86
        sympathetic branch, 84–86
    axon reflex, 83, 88
    dorsal root ganglia, 83–84
    enteric ganglia, 86
    enteric nervous system, 82–83, 85–86
    mesenteric/coeliac ganglia, 86
    myenteric plexus, 86
    paravertebral ganglia, 85–86
    postganglionic neurons, 85–86
    preganglionic neurons, 85
    prevertebral ganglia, 85–86
    somatosensory pathways, 83–84
    splanchnic nerves, 84
    splenic nerve, 86
    submucosal plexus, 86
    trigeminal nerve, 84

    vagus nerve, 84, 89
    visceral afferent pathways, 82–83, 84–90
Peripheral neuropathy, 187
Peripheral tolerance, 164–165
Perivascular macrophages, 182
Personal achievement, 107
Personality, 99–101, 105, 119–120, 141–142, 150, 170–171, 188
Pervasive developmental disorders (PDD), 196
Pessimistic, 163
Peyer's patches, 50
Phagocytes, 16, 53–54, 128, 166–167, 182, 218
Phagocytosis, 28, 71, 89, 124, 216, 219
Pharmaceuticals, 144, 230
Pharyngitis, 137
Phenotype, 162
Photophobia, 187
Physical activity, 171, 215, 221, 228
Physical contact, 228
Physical exercise, 208, 221
Physical exertion, 195
Physical trauma, 103, 105, 215
Physiochemical properties, 230–231
Physiological autoimmunity, 164
Phytohemagglutinin (PHA), 52
Pineal gland, 62, 73, 75–76, 78
Pitta, 13
Pituiary-adrenal axis, 67, 120–121, 226
Pituitary cells, 58, 67
Pituitary adenylate cyclase activating polypeptide (PACAP), 88
Pituitary-gonadal axis, 121
Pituitary stalk, 60
Pituitary-thyroid axis, 121
Placebo, 4, 139, 148, 150, 212–213, 232
Placebo effect, 209–210, 221
Placenta, 68, 70
Plaque of Athens, 134
Plasma cells, 40, 125
Plasma proteases, 29
Plasmapheresis, 188
Platelet activating factor, 28–29
Platelets, 146
Pleiotropy, 45
Pneumocystis carinii, 137
Poison ivy, 143
Pokeweed mitogen (PWM), 52
Poliomyelitis, 138
Pollen, 143
Polymorphonuclear leukocytes, 25

Poor concentration, 111
Portal capillaries, 59
Postencephalitic behavior, 197
Posterior pituitary, 59, 73
Postganglionic neurons, 75, 85–86
Postpartum period, 169, 172
Potions, 11
Prayers, 11
Preganglionic autonomic neurons, 85
Pregnancy, 71, 78, 107, 169–170, 172, 173,
    175–176, 188, 201, 226
Premenstrual phase, 169
Preoptic area, 65
Prevention, 230
Prevertebral ganglia, 85–86
Primary lymphoid organs, 46, 49, 82, 86, 89
Primates, 120–121
Prions, 136, 184, 186
Privacy, 231
Procaryotes, 136
Procreation, 226
Profit, 233
Progesterone, 62, 70–71, 169, 174–175
Proinflammatory cytokines, 45, 90, 97, 172, 175,
    182, 190, 195–196, 209, 217
Proinflammatory mediators, 89, 145, 147
Prolactin (PRL), 60, 63–64, 66–67, 70, 74,
    76–77, 121, 129–130
Pro-opiomelanocortin (POMC) protein, 61,
    76
Prospective studies, 139, 174, 231
Prostaglandin-E_1, 75
Prostaglandin-E_2, 75, 192, 220
Prostaglandins, 28–29, 69, 145, 220
Protein, 34, 60, 136, 144, 146, 167–169, 183,
    186, 190, 199, 218, 220
Protozoa, 137
Protozoan diseases, 137
Psoriasis, 209
Psychiatric disorders, 19, 150, 168, 173, 180, 186,
    188, 190–191, 196, 203, 227, 232
Psychodynamic, 148
Psychological state, 84
Psychologic challenges, 195
Psychoneuroimmunology, 1, 4, 5, 7, 18, 20,
    225–226, 228–230, 232–233
  healthcare implications, 232
  historical antecedents, 18–19
  research implications, 229–232
Psychosocial dwarfism, 121

Psychosocial stress, 3, 118–119, 123, 134, 139,
    141–142, 148–149, 151, 154, 159–160,
    162–163, 170, 173–176, 203, 208, 211, 215,
    217, 221, 225, 227
  academic exams, 118, 124–125, 127, 130
  contextual change, 3, 103–115, 129
  earthquake, 118, 124, 127–128, 130
  electric shock, 2, 119, 124
  laboratory induction, 119
  life events, 106, 118, 140, 142, 150, 159, 161,
      163, 170, 175, 188
  mental activity, 108, 113, 127
  mental arithmetic, 113, 119, 124
  paratrooper training, 118, 121–122
  public performance, 118
  public speaking, 113, 119, 124, 127
  rapid decisions, 119
  role of personality, 119
  social environment, 109, 118, 126
  sporting events, 118
  war, 118, 124
Psychosocial stress effects on
  acetylcholine, 122
  acquired immune deficiency syndrome,
      140–143
  arginine vasopressin, 121
  asthma, 148–149
  breast cancer, 161–162
  CD4 to CD8-ratio, 126–127
  cervical cancer, 163
  colon cancer, 160
  cytokine profile, 127
  cytomegalovirus antibody titers, 126
  delayed-type hypersensitivity reaction,
      126–127
  enkephalins, 122
  epinephrine, 122
  Epstein-Barr virus antibody titers, 126–127
  gabaergic neurons, 123
  Graves disease, 170
  growth hormone release, 121
  herpes simplex virus antibody titers,
      126–127
  infectious mononucleosis, 140
  inflammatory-bowel disease, 173
  influenza, 140
  insulin dependent diabetes, 173
  leukocyte count, 124–129
  B-lymphocyte count, 125
  T-lymphocyte count, 126–127

melanoma, 161–162
mitogen induced proliferation, 126–127
monoaminergic neurons, 123
multiple sclerosis, 174
natural killer cell activity, 124–125
norepinephrine, 122
oxytocin release, 121
pancreatic cancer, 160
phagocytosis, 124
pituitary-adrenal axis, 120–121
pituitary-gonadal axis, 121
pituitary-thyroid axis, 121
prolactin release, 121
rheumatoid arthritis, 170–172
secretory IgA, 125–126
serotonin, 123
substance P, 122
systemic lupus erythermatosus, 173–174
upper respiratory infections, 140–143
Psychosocial treatment, 161–162
Psychosomatic disorders, 99, 148, 150, 170
Psychosomatic medicine, 18, 148
Psychotic disorders, 187, 191, 199, 201–202
Public health, 135, 149, 231, 233
Public Health Act of 1848, 15
Public performances, 118
Public speaking, 113, 119, 124, 127
Putamen, 97
Puzzle solving, 124

Q
Qi, 12
Qiqing, 12

R
Rabies, 19, 189, 198
Rabies virus, 181, 189
Radioactive chromium, 52
Radioactivity, 157
Radiation, 103, 110–111, 114, 157, 230
Rajas, 13
Raphe nuclei, 95
RAS gene, 156
Rasmussen, 18
Razi, 171
Reactive oxygen species, 219
α-Receptors, 74
β-Receptors, 74–75
Receptors, 114
Recombinase machinery, 37

Reductionistic approaches, 229, 233
Relationships, 109, 140, 171, 173, 209
Relaxation, 4, 214–215, 221, 228
Relaxation habits, 208
Releasing hormones, 59
Religions traditions, 208
Remedy, 208
Renaissance, 11, 15
Reproductive tract, 68
Research, 225, 229, 231
Resilience, 115
Resistance phase, 104
Respectful, 163
Retrospective studies, 140, 170
Reticular formation, 84, 94–95
Retina, 75
Retirement, 106–107
Retroviral genomes, 201
Retroviruses, 111, 136
Reye's syndrome, 185
Rheumatic fever, 137–138, 145, 167, 199
Rheumatoid arthritis, 145, 165, 167, 170–173,
    176, 195, 202, 214
Rheumatoid factor (RF), 172
Rhinitis, 143
Rickettsias, 136
Right-hemisphere, 96–97, 214, 221
Rigid, 140
Rimon, 171–172
Rituals, 11
RNA, 136, 188
RNA editing, 36
Roundworms, 137
RU 486, 175
Rubella virus, 196, 201

S
Sankhya, 12
Sattva, 13
Scabies mite, 137
Schizophrenia, 4, 19, 148, 160, 173, 187–188,
    199–203, 232
Schwann cells, 188
Science, 233
Scientific methodology, 229
Scleroderma, 165
Season of birth, 200
Secondary lymphoid tissue, 33, 47, 49, 82, 91
Secrecy, 232
Secretory IgA (sIgA), 50, 125, 210, 215–217

Seizures, 185, 188, 191, 203
Selectins, 46
Selenium, 219
Self/nonself discrimination, 24, 226
Self-sacrifice, 171
Self-similarity, 229
Selye, Hans, 18, 69, 103–108, 114, 194
Seminaria, 134
Sensitization, 210
Separation, 105, 171, 173
Sepsis, 218
Septal area, 60, 95
Septic shock, 45
Seropostive, 172
Serotonergic receptors, 87
Serotonin (5-HT), 28–29, 68, 95, 123
Shock reaction, 104
Sickness behavior, 4, 94, 96, 193–196
Side chains, 16
Sjögren's syndrome, 165, 195
Skin, 68, 82–83, 165, 167–168
Sleep, 4, 142, 194–195, 214, 221
Sleep deprivation, 124
Sleeping sickness, 137
Sleeplessness, 186
Slow wave sleep, 214
Small-cell lung cancer, 168
Smallpox, 134
Social activity, 208, 213, 231
Social context, 109, 118
Social dominance, 120
Social engagement, 213–214, 221, 228
Social experience, 228
Social institutions, 105
Social network diversity, 142
Social networks, 3, 108
Social setting, 106
Social support, 126, 140, 150, 161–162, 175, 213
Social withdrawal, 200
Socioeconomic status (SES), 139, 172
Solomon, George Freeman, 18–19, 203
Somatic mutations, 157
Somatosensory axons, 84
Somatostatin (SS), 29, 60, 65, 74, 88, 90
Somatotrophs, 64
Somatotropin release inhibiting hormone (SRiH), 60
Soybeans, 143
Spaceflight, 125, 127–128, 130
Specific antibody titers

Specific genes, 230
Specific nutrients, 217
Spiegel, David, 161–162
Spinal cord, 60, 78, 92, 185
Spirochetes, 136–137
Splanchnic nerves, 84
Spleen, 33, 47–49, 72, 86–87, 92, 128, 167
Splenic nerve, 86
S. pneumoniae, 184
Spondyloarthropathies, 168
Sporting events, 118
Starling, 17
Starvation, 218
Sternberg, Esther, 17
Stevens-Johnson syndrome, 144
Stimulant drug, 212
Stimulating hormones, 59
Stoic style, 161
Strep-throat, 138
Streptococcus pyogenes, 137
Stress, 2, 78, 103–115, 117, 124–126, 130–141,
    149–150, 163, 170–171, 173, 176, 188,
    194–195, 201, 213, 215, 217–218, 220–222,
    227
Stress management, 213
Striatum, 198
Stroke, 96
Stromal cells, 43
Subacute encephalitis, 184
Subfornical organ (SFO), 60, 73, 88, 94
Submucosal plexus, 86
Substance P (SP), 29, 68, 85, 88, 90, 122, 147
Substance use, 142
Substantia nigra, 95
Suggestion, 3, 148, 150, 208–210
Suicide, 160
Super antigens, 185
Super oxide anion (O_2^-), 66
Superior cervical ganglion, 75
Suprachiasmatic nucleus, 75
Surgery, 127, 160, 170, 188
Swedo, 198
Sydenham's chorea, 185, 198
Syphilis, 194
Sympathetic nervous system (SNS), 17, 49, 74,
    82, 85–86, 90, 129, 215
Systemic autoimmunity, 167
Systemic lupus erythematosus (SLE), 165, 167,
    173, 176, 186, 190, 195
Systemic mycoses, 137

T

Tachycardia, 145
Tachykinins, 147
Tamas, 13
Taoist, 12
T-cell count, 141, 202
T-cell proliferation, 75, 89, 127, 216
T-cell receptor (TCR), 37–38, 226
T-cell surface molecules
Team approach, 232
Technology, 108, 157
Temperament, 100
Temporal lobes, 188
Testicular cancer, 168, 191
Testes, 62, 70
Testosterone, 62, 65, 70–71, 121, 129
Tetanus toxin, 185
Th 0, 39
Th 1, 39, 67, 69, 71, 90–91, 127, 145, 166, 174, 202, 219
Th 2, 39, 69, 71, 91, 128, 145–146, 148, 166, 173, 202
Thalamus, 60, 92–93
Thoracic duct, 47
Thorazine, 191
Thymic peptides, 51, 77
Thymocytes, 76–77, 91, 95
Thymopoietin, 63, 76
Thymosin, 63, 76
Thymosin β4, 70, 76
Thymosin-α1, 76
Thymulin, 63, 72, 76–77
Thymus, 46–47, 49, 51, 63, 68, 70, 72–73, 76–77, 86
Thyroid disorders, 170, 176
Thyroid stimulating hormone (TSH), 60, 63–64, 72, 76, 121
Thyrotrophs, 72
Thyrotropin releasing hormone (TRH), 60, 72
Thyroxine (T$_4$), 62, 72
T-Lymphocytes, 25, 31, 33, 43–45, 49–50, 53–54, 65–66, 75, 89, 96, 138, 164–165, 169, 174, 176, 182, 185, 196, 209, 214, 218
Tobacco, 157, 162, 195
Tobacco smoke, 145, 175, 220
Tolerance, 37, 39, 78, 146, 164, 175, 226
Tourette's syndrome, 198–199
Toxic chemicals, 114, 157, 168, 176, 203, 228
Toxicity, 105

Toxoplasmosis, 137
Transforming growth factor-β (TGF-β), 44, 182–183
Traumatic experiences, 170
Treatment, 230
Treponema pallidum, 187
Trigeminal, 84
Triiodothyronine (T$_3$), 62, 72, 77
Tritiated-thymidine, 52
Trojan horse mechanism, 181
Tsetse fly, 137
Tuberculin test, 52, 209
Tuberoinfundibular region, 66
Tumor, 154, 159–160, 162–163, 168, 191, 212
Tumor necrosis factor (TNF), 30, 39, 45, 68–69, 90–91, 138, 166, 169, 182, 193, 217, 220
Tumor specific antigens, 158, 175
Tumor suppressor genes, 156, 175
Tumoricidal activity, 68
Twins, 144, 174

U

Ulcerative colitis (UC), 173
Ultrasound stimulation, 212
Ultraviolet (UV) radiation, 145, 220
Unconditioned response (UCR), 98–99
Unconditioned stimulus (UCS), 98–99
Unemployed, 127
Unethical practices, 231–232
Unicellular organisms, 136
Upper respiratory viral infection, 140, 142, 215
Urticaria, 143, 209
Uveitis, 165

V

Vacation, 105, 107
Vaccination, 19, 126, 135, 188
Vagus nerve, 73, 84, 89, 193, 196
Van Leeuwenhoek, Anton, 15
Variable segment, 36
Varicella-zoster, 187, 210
Vascular endothelial cells
Vasculitis, 190
Vasoactive intestinal peptide (VIP), 20, 66, 85, 88, 90–91, 147
Vãta, 13
Vectors, 108, 134–135, 187
Veiled cells, 30

Ventral medulla, 92
Ventromedial nucleus of the hypothalamus, 96
Ventromedial prefrontal cortex, 94
Verbal communication, 214, 228
Vesalius, Andreas, 15
Viral replication, 126
Viremia, 181, 189
Virion, 136
Viroids, 136
Virulence, 135
Viruses, 53, 111, 136, 140, 158, 181, 183–184, 186, 189, 201
Visceral afferent pathways, 82, 84–90
Vitamin A, 219
Vitamin C, 219
Vitamin D, 219
Vitamin E, 219
Vitamins, 218–219
Volunteers, 119, 142
Von Pirquet, 143
Voodoo death, 211

W

Wakefield, 197
Waldeyer, Wilhelm, 155
War, 118, 124
Warts, 209

Water, 110, 230
Weather, 108, 110
Weber and Weber, 17
White blood cells (WBC), 69, 124, 195, 215
Willis, 17
Winslow, Jacobus, 17
Wisdom teeth extraction, 212
Worry, 124
Wright, Almroth, 16

X

Xin, 12

Y

Yeasts, 137
Yellow fever, 110
Yernisia Pestis, 134
Yernisia enterocolitis, 137
Yin-yang, 12
Yoga, 217
Yolken and Torrey, 203

Z

Zhengqi, 12
Zinc, 218–219
Zoonotic infection, 134